SpringerBriefs in Applied Sciences and Technology

More information about this series at http://www.springer.com/series/8884

Gernot Stoeglehner · Georg Neugebauer
Susanna Erker · Michael Narodoslawsky

Integrated Spatial and Energy Planning

Supporting Climate Protection and the Energy Turn with Means of Spatial Planning

Gernot Stoeglehner
University of Natural Resources
　and Life Sciences
Vienna
Austria

Georg Neugebauer
University of Natural Resources
　and Life Sciences
Vienna
Austria

Susanna Erker
University of Natural Resources
　and Life Sciences
Vienna
Austria

Michael Narodoslawsky
Graz University of Technology
Graz
Austria

ISSN 2191-530X   ISSN 2191-5318 (electronic)
SpringerBriefs in Applied Sciences and Technology
ISBN 978-3-319-31868-4   ISBN 978-3-319-31870-7 (eBook)
DOI 10.1007/978-3-319-31870-7

Library of Congress Control Number: 2016935211

© The Author(s) 2016
This work is subject to copyright. All rights are reserved by the Publisher, whether the whole or part of the material is concerned, specifically the rights of translation, reprinting, reuse of illustrations, recitation, broadcasting, reproduction on microfilms or in any other physical way, and transmission or information storage and retrieval, electronic adaptation, computer software, or by similar or dissimilar methodology now known or hereafter developed.
The use of general descriptive names, registered names, trademarks, service marks, etc. in this publication does not imply, even in the absence of a specific statement, that such names are exempt from the relevant protective laws and regulations and therefore free for general use.
The publisher, the authors and the editors are safe to assume that the advice and information in this book are believed to be true and accurate at the date of publication. Neither the publisher nor the authors or the editors give a warranty, express or implied, with respect to the material contained herein or for any errors or omissions that may have been made.

Printed on acid-free paper

This Springer imprint is published by Springer Nature
The registered company is Springer International Publishing AG Switzerland

Preface

Energy issues and spatial planning become increasingly interwoven. The more we talk about sustainable energy systems, energy resource limitations or greenhouse gas reduction, and climate change mitigation, the more we move into the thicket of the jungle of interrelations between space and energy. This book is meant to provide the reader with a reliable compass but also with other necessary tools to find his way through this jungle and arrive at the other side of the thicket at practical planning solutions.

One of the vexing features of the link between energy and spatial planning is that it cuts not only across two of the most basic scientific concepts, space and energy, but puts very different actors with various backgrounds in the same boat. This is true on the scientific level, where integrated spatial and energy planning is almost synonymous with interdisciplinary. It requires expertise in spatial planning, energy, and grid engineering, but also in fields as far apart as agriculture and civil engineering. This variety of actors applies however to the planning process, too. Gone will be the times where the only principal players for energy systems will be big utilities, public authorities, and possibly developers and investors. In an area of dynamic technological progress in both energy provision and energy efficiency technologies, with ever-increasing possibilities for decentralized small-scale installations and "smart grids," energy systems can become democratized. This makes planning processes arguably more complex as they become increasingly participatory. It makes them however certainly more interesting and challenging.

This increased complexity of the spatial planning process when energy issues become involved also requires a new quality for any tool used in this endeavor. Tools become means to facilitate the discourse between stakeholders. They have to provide scientific and technical rigor while allowing at the same time to accommodate different viewpoints and interests of stakeholders. It is not only the end result of a calculation or design that counts. Almost as important is the information about different aspects of complex spatial solutions for energy systems that stakeholders can glean from using a tool. Tools must make hidden systemic links within energy systems transparent. They must also provide a graphic and understandable description of the behavior of energy systems within spatial contexts, so that stakeholders can base their decisions on their outcome. This means that as integrated spatial and energy planning becomes more systemic, so must the tools involved.

According to the systemic and interdisciplinary challenges of integrated spatial and energy planning, this book has been an interdisciplinary exercise. It combined authors with spatial planning as well as engineering backgrounds who had had already a long tradition in jointly working in integrated spatial and energy planning projects in various settings. Besides providing different vantage points on integrated spatial and energy planning and fuse engineering as well as spatial planning expertise based on long cooperation, it is the practical experience in using tools in participatory planning processes that is important. The book is meant to provide practical guidance for integrated spatial and energy planning that requires guides who do have theoretical knowledge but can also report from experience.

The authors are therefore particularly grateful to those persons and institutions that allowed them to gain this experience in a number of interesting and challenging action research projects and for providing valuable feedback on the application of the tools described in this book. We want to thank the Austrian Climate and Energy Fund for supporting fundamental as well as applied research projects such as the project PlanVision and the development of the ELAS Calculator, as well as the establishment of the resource plan for the Mühlviertel region. The city of Freistadt has been a particularly valuable partner in many projects, some of them co-funded by the Austrian Climate and Energy Fund. Without that support and cooperation neither the Energy Zone Mapping could have been developed nor could the progress in Process Network Synthesis have been achieved that finally led to the development of RegiOpt.

In many projects Austrian Federal State administrations were either funding partners or valuable sounding boards. We are particularly thankful to the Federal Ministry of Agriculture, Forestry, Environment, and Water Management as well as to the Federal State Government of Upper Austria, Salzburg and Lower Austria for their support. Our thanks also extend to the cities of Graz and Vienna for providing support, data, and funds for projects that considerably contributed to honing the tools presented here. Especially, we want to thank our action research and practice partners with whom we have established long-lasting cooperation in projects that constitute the basis of this book, in particular (in alphabetical order) Hans Emrich, Winfried Ginzinger, Christian Jachs, Siegfried Kautz, Helmut Koch, Gilbert Pomaroli, Friedrich Stockinger, Oskar Stöglehner, and Werner Thalhammer. Finally, we want to thank our fellow researchers in these interdisciplinary research projects.

We hope that this book will help the reader to get new insight into integrated spatial and energy planning. We also hope that the theoretical framing, the planning principles, and the tools presented here will prove helpful in the challenge to create sustainable solutions that reduce the environmental impact from energy provision and provide secure and sufficient energy supply while reducing the long-term costs for energy services and infrastructure.

Vienna, Graz
May 2016

Gernot Stoeglehner
Michael Narodoslawsky
Susanna Erker
Georg Neugebauer

Contents

1	**Introduction**	1
	1.1 Energy Systems and the Energy Turn	2
	1.2 Spatial Structures and the Energy Turn	5
	1.3 Scope of the Book	7
	References	8
2	**System Interrelations Between Spatial Structures, Energy Demand, and Energy Supply**	11
	2.1 Energy Efficiency and Spatial Structures	14
	2.2 Renewable Energy Provision and Spatial Structures	21
	2.3 Energy Logistics and Spatial Structures	22
	2.4 Energy Resilience and Spatial Structures	27
	References	31
3	**Spatial Archetypes in the Energy Turn**	35
	3.1 Urban Areas	38
	3.2 Rural Areas	41
	3.3 Rural Small Towns	43
	3.4 Suburban Areas	45
	3.5 Suburban Small Towns	47
	3.6 Mix of Spatial Archetypes	50
	References	51
4	**Fields of Action for Integrated Spatial and Energy Planning**	55
	4.1 Energy-Efficient Spatial Structures	60
	4.2 Renewable Resources and Spatial Structures	65
	4.3 Energy Supply Systems Tailored to Spatial Structures	68
	References	71

5	Measures for Integrated Spatial and Energy Planning		73
	5.1	Integrated Spatial and Energy Plans on the Regional Scale	73
		5.1.1 Definition of Core Areas for Regional Spatial Development	73
		5.1.2 Integrated Resource and Energy Concepts	74
	5.2	Integrated Spatial and Energy Plans on the Local Scale	76
		5.2.1 Measures Concerning Building Land	76
		5.2.2 Measures Concerning Open Space	78
		5.2.3 Measures Concerning Infrastructure and Mobility	79
		5.2.4 Measures Concerning Energy Supplies	79
	5.3	Measures for Existing Spatial Structures	80
		5.3.1 Residential Areas with Apartment Buildings and/or Mix of Functions	81
		5.3.2 Single-Family Housing Areas	82
		5.3.3 Mixed-Function Areas in Central Locations	83
		5.3.4 Industrial and Commercial Areas	84
		5.3.5 Shopping Centers	86
		5.3.6 Outer Rural Areas	88
	5.4	Concluding Remarks	89
	References		90
6	Processes and Tools for Integrated Spatial and Energy Planning		91
	6.1	Theory Framework	91
	6.2	Top-Down Framework Planning and Bottom-up Action Planning	94
	6.3	Tools for Integrated Spatial and Energy Planning	97
		6.3.1 Elas	99
		6.3.2 Energy Pass for Settlements 2.0	101
		6.3.3 Energy Zone Mapping	104
		6.3.4 RegiOpt	106
	6.4	Concluding Remarks	108
	References		109
7	Résumé		111

List of Figures

Figure 2.1	System analysis spatial and energy planning (own illustration after Stoeglehner et al. 2011b)..........	13
Figure 2.2	Energy aspects of residential settlements (own illustration after Stoeglehner et al. 2014a)..........	15
Figure 2.3	Energy cascade (own illustration)....................	18
Figure 2.4	The influence of siting on energy demand (own illustration after EnergieAgentur.NRW 2008; Treberspurg 1999)...............................	19
Figure 2.5	Energy-harvesting density for different energy resources (Narodoslawsky 2014).....................	22
Figure 2.6	Cost for electricity storage (Narodoslawsky 2014). Methane refers to methane generated by power-to-gas processes, Hydrogen produced by electrolysis, CAES Compressed Air Energy Storage, PSP Pump Storage of Power, Biogas produced from grass silage...........	26
Figure 2.7	Principles of spatial energy resilience (own illustration)................................	29
Figure 3.1	Classification of the Austrian territory according to five spatial archetypes (own illustration after Stoeglehner et al. 2011; Exner et al. 2016)............	38
Figure 4.1	Influencing factors for integrated spatial and energy planning (own illustration after Stoeglehner et al. 2014)...........	56
Figure 4.2	Actors and stakeholders relevant for integrated spatial and energy planning (own illustration after Stoeglehner et al. 2014).........................	58
Figure 4.3	The "decision-making pyramid" (own illustration after Stoeglehner and Narodoslawsky 2008)............	59
Figure 6.1	Single- and double-loop learning (own illustration after Argyris 1993; Innes and Booher 2000; Stoeglehner 2010)...............................	92

Figure 6.2	Double-loop learning with strategic planning and assessment methods (own illustration after Stoeglehner 2014).	94
Figure 6.3	The indicator pyramid (own illustration after Stoeglehner and Narodoslawsky 2008; Stöglehner 2014).	98
Figure 6.4	The energy demand of a residential settlement as one part of the ELAS results (Stoeglehner et al. 2011a)	100
Figure 6.5	The final rating of the Energy Pass 2.0 (Emrich et al. 2012).	103
Figure 6.6	Energy Zone Mapping as a basis for the local development concept of an Austrian urban settlement (own illustration after Stoeglehner et al. 2011b).	105
Figure 6.7	Ecological comparison of an optimal structure with business as usual, using the SPI—ecological footprint (own illustration after RegiOpt 2016)	107
Figure 7.1	A generic scheme of energy-optimized local spatial structures (own illustration).	113

Chapter 1
Introduction

Gernot Stoeglehner, Michael Narodoslawsky, Susanna Erker, and Georg Neugebauer

Abstract "Integrated spatial and energy planning" constitutes an important strategy to implement the energy turn. This chapter gives a brief introduction, discussing the dynamic of energy flows as well as the spatial distribution of energy sources and energy demand first. Secondly, spatial structures determine the possibilities to support the energy turn in given spatial contexts and how energy efficiency can be organized on the scale of spatial structures. Finally, the scope of this book is defined.

Shifting from a fossil and nuclear to a renewable energy resource base is an imperative societal and political target. Numerous political strategies have been formulated in recent years from the supranational, national to the regional and local levels to move toward this "energy turn."[1] Among others, arguments for the energy turn are climate change mitigation (IPCC 2014), the anticipated scarcity of fossil and nuclear resources in the near future (catchword "peak oil," see, e.g., Aleklett 2012), the use of regional energy resources in order to gain more autonomy from fuel imports, the striving for positive regional–economic effects, or the creation of green jobs. Many strategies for the long-term implementation of the energy turn have been formulated that mainly rest on two pillars (see, e.g., European Commission 2011a; BFE 2015; BMWi 2015; BMLFUW and BMWFW 2010): (1) the reduction of the energy consumption, e.g., by changes of lifestyles and economic practices as well as energy efficiency measures; and (2) the substitution of fossil and nuclear energy by renewable energy sources.

In policy making, research, and development, much attention has been paid to the technological aspects of the energy turn, which allows us now to choose between a wide range of options for energy saving and the generation of renewable energy, with some technological issues such as energy storage still partly unsolved. Yet, implementing the energy turn does not only mean to deal with technologies. A complex fabric of issues influences possibilities and options to proceed toward the energy turn, which are, inter alia, the base values of society, the interplay of

---

[1]In the German-speaking world, "Energiewende," which literally translates to "energy turn," is the common phrase to describe this shift toward a renewable energy system.

different policies with relevance for energy policy (e.g., economic policies, agricultural policies, fiscal policies, environmental policies), the availability of technologies, regional and local resource potentials, demographic development of societies, individual lifestyles, economic practices as well as the physical and planned spatial development (Stoeglehner et al. 2014a). The list can be further expanded and deepened with the issues of educational backgrounds and awareness of populations which have a direct effect on the base values of societies, or societies' and decision makers' capacity to learn about alternative options to reach the energy turn, adding to the complexity of the problem at hand.

In this book, we focus our attention on the spatial aspects of the energy turn. Concerning both energy saving and the extended use of renewable energy sources, the organization of spatial structures heavily influences the technological and economic options to conceptualize renewable-based energy systems. This fact is normally not paid enough attention to in strategy formation for the energy turn, but if spatial structures and spatial development are taken into consideration, the quality of planning can be increased a lot (Stoeglehner et al. 2011b). Furthermore, also the acceptance of measures by local populations differs in various spatial contexts. Therefore, based on our research work carried out since 2007 (Erker et al. 2015; Exner et al. 2016; Regionalmanagement Oberösterreich 2012; Neugebauer et al. 2015; Stoeglehner et al. 2010; 2011a, b, c; 2014a, b; 2015), we propose that "integrated spatial and energy planning" should be an important part of any holistic strategy to reach the energy turn targets. Integrated spatial and energy planning can be defined as the part of spatial planning that deals with the spatial dimensions of energy consumption and energy supply (Stoeglehner et al. 2014a). This book clarifies what these spatial dimensions are. Therefore, we emphasize on systems, fundamental system interrelations, and conceptual frameworks.

1.1 Energy Systems and the Energy Turn

We argue that the concepts of *space, time,* and *energy* constitute a significant basis of our understanding of the physical reality. None of these concepts is viable without the others as they are inherently interwoven. Of these concepts, energy is arguably the least understood. Therefore, it is necessary to explain some features of this eluding concept and its relation to the others—time and space.

The first interesting feature of energy is that we actually do not need it. We need energy services (see, e.g., WIFO 2011) such as comfortable residential climate, bright spaces in the night, communication, social, and cultural interaction. In providing all these services, energy is only one of many factors that may be substituted by each other to a certain extent, as can be explained with comfortable warm rooms when ambient temperatures become chill. The site of the building, architectural and technical features of the building, the heating system, and even the behavior of residents—let alone their expectations—contribute to how comfortable a room is in wintertime. Choosing an appropriate site and furnish the house according to passive house standards may be one way to accomplish warm rooms. Another might be to install a powerful heating system. It is obvious that energy provision is less of a factor in the first case and dominant in the second.

A second interesting feature of energy is that it can neither be produced nor be destroyed. Energy services are provided by converting energy between different forms. Three levels of energy quality can be distinguished: high-quality energy such as electricity, mechanical energy, or high-temperature heat (above 100 °C); medium-quality energy such as medium-temperature residential heat, or process heat (50–100 °C); and low-quality energy such as ambient heat, residential heat and process heat, off-heat (with less than 50 °C). This means that the provision of energy always requires energy flows, i.e., energy per time unit, that are converted from one energy form into another. The laws of thermodynamics govern the conversion of energy (Bailyn 1994). While the first law of thermodynamic defines that energy may neither be produced nor be destroyed, the second law of thermodynamic rules quality of energy flows subjected to conversion. Without going into any details, the general rules imply the following aspects for energy system design:

- All energy that flows into a system must either be stored in the system or left, possibly in another form and quality (first law).
- High-quality energy can be converted into lower quality energy forms (e.g., heat) with almost no losses; this is at the basis of all heating processes, where high-quality energy (e.g., electricity, high-temperature heat) is converted to lower quality energy (process heat, residential heat).
- Lower quality energy can be split into a part of higher quality and another part of lower quality energy; this is at the basis of the concept of combustion engines that always transform high-temperature heat into high-quality energy (mechanical energy, electricity) and low-quality energy (off-heat).
- Low-quality energy (e.g., ambient heat) can be upgraded to medium-quality energy (e.g., residential heat or process heat) by the infusion of high-quality energy (electricity, high-temperature heat). This is the principle of heat pumps. The fraction of high-quality energy necessary to upgrade low-quality energy rises, the larger the temperature difference between low and medium level becomes.

A third interesting aspect of energy concerns the dynamic of energy flows. As stated above, energy flows entering a system must either leave it or be stored in it. Energy services are always linked to the conversion of these flows. Social and economic processes govern the dynamics of energy services demand. Natural energy flows, in particular solar radiation, wind, and water flows, are determined by the dynamic of natural processes. Whenever energy service demand and the provision of energy flows do not match in temporal terms, either storage elements are necessary within the system or the intake or delivery of energy to connected energy systems (Ramirez et al. 2015). Energy forms differ quite substantially in their difficulty and cost of storage (eseia 2014): Material energy sources such as crude oil, natural gas, biomass, and hydrogen (which contain energy in the relatively high quality of chemical energy) can be stored relatively easy with almost no loss. Storing heat is also quite easy and cheap however burdened with losses. These losses are higher, the higher the quality of heat is, i.e., the higher the temperature of the medium to be stored. The situation is different for storing electricity. This energy form must be transformed to other forms such as chemical energy in batteries and power-to-gas (P2G) systems or potential energy in hydro-pump storages. From these energy forms, electricity may then be retrieved when it is necessary to provide energy services. This transformation process is always complex and costly.

Finally, the spatial distribution of energy sources and energy demand can show significant, but different disparities, so that matching the areas or origin of energy sources and the locations of energy services demand becomes a logistics problem. Fossil fuels are available from limited areas around the globe, being often transported for several thousand kilometers before they reach the points of conversion and consumption. This can be also true for certain large-scale renewable energy sources such as wind power or large-scale PV systems, where electricity is transported over hundreds or thousands of kilometers with respective losses. Contrarily, solar radiation as the basic source of all renewable energy forms, as well as biomass energy, are distinctly decentralized resources, whereas energy service demand is mainly concentrated in urban and industrial areas. This requires an intricate logistical system to bring energy carriers and available energy flows to the places where they may be converted to provide energy services. Again, different energy forms vary widely in their logistical parameters and in the costs, losses and necessary infrastructure to transport them. Furthermore, energy has to be collected from huge areas, and energy sources have to be converted into transportable forms, so that especially renewable energy supplies reveal the interlinkages of urban centers with their supply hinterland.

The flows of energy sources through the logistics systems to the points of use where conversion technologies are linked with other factors to render the required energy services constitute highly complex systems. Changes in one element, say the dominant source of energy flows, will have profound impact on the system setup and all other elements. The energy turn as a total shift of the resource system from fossil and nuclear energy to renewable sources will have dramatic implications for this energy system.

Besides a change in the sources, the energy turn also requires avoiding and restricting adverse effects of the whole energy system on our natural environment. It is important to note that changing the resource base goes a long way toward this goal but is by no means sufficient. Even utilizing renewable energy sources may be done in ways that are by no means sustainable. Sustainable energy systems thus do not only imply a reorientation toward renewable energy sources, but they also have to utilize natural resources according to the principles of sustainability.

Many features of the change in the energy system mandated by the energy turn have spatial connotations, including the demand for energy services, the harvesting of decentral renewable energy sources; the logistics of guiding renewable energies and energy sources to the areas where energy services are needed; and the spatial distribution of optimal energy storage and conversion technologies. This also applies to the role that energy plays in providing the services demanded by society, which is strongly influenced by spatial structure and operation of infrastructure and even more so by societal expectations and behaviors.

1.2 Spatial Structures and the Energy Turn

Appraising the relevance of spatial structures for the energy turn, three main features of spatial structures are of most importance (Stoeglehner et al. 2011b): their mix of spatial functions as well as their density and size.

The mix of functions relates to the fundamental spatial functions, which are dwelling, working, food provision, recreation, supply, and disposal as well as communication and mobility between the other five (Lienau 1995). If the mix of functions is high, an attractive amount of offers for dwelling sites, work places, recreation and shopping facilities, transport options, etc., (Heinze and Kill 1995) is available for the inhabitants of such structures. Mix of functions also forms the basis for walking and biking, as the distances between the functions of daily life can be kept short (Prehal and Poppe 2003). Taking a holistic picture into account, further functions such as nature protection and environmental compensation for societal pressures have to be taken into consideration. Therefore, a mix of functions is promoted in many planning visions both on the regional and on the local scale mainly for the increase of quality of life for the inhabitants (CNU 2001; Dittmar and Ohland 2004; European Comission 2011b; Farr 2008; Gaffron et al. 2005, 2008; Newman and Jennings 2008; Prehal and Poppe 2003; Register 2002; Schriefl et al. 2009; UNEP 2002). Mix of functions reduces travel times and distances and obviously also saves energy. But it also stands for a more efficient supply with grid-bound energy (Stoeglehner et al. 2011c).

Density is a measure of intensity and efficiency. In spatial planning, density normally increases the cost efficiency of public infrastructures, saves unsealed soil, and is also discussed in light of energy and resource efficiency. Yet, density is ambivalent: On the one hand, if density is low, also cost efficiency is low and environmental pressure is high. Suburbanization is, inter alia, often criticized for too low densities (Stoeglehner et al. 2011b). On the other hand, if density is too high in urban areas, quality of life might suffer, so that certain planning visions such as the broadacre city were explicitly developed as a model against urban density (Magnago Lampognani 2010). Therefore, density has to be cautiously considered, taking both issues of quality of life as well as efficiency into account.

The same is true for size, which is closely related to density and mix of functions: if a certain land use's size increases too much, distances to the next function or to specific subcategories within the same function become to big, so that certain efficiency effects are overcompensated by a higher transport demand. Therefore, when it comes to resource use, an economy of scale—the bigger the better—is complemented by an ecology of scale—meaning that the smaller the better

(Gwehenberger et al. 2007). Concerning renewable energy provision, the size and density of spatial structures determines their energy intensity and their supply hinterland if the total energy demand cannot be covered on-site.

Several recent planning visions for spatial development on the urban and regional scale point in the same direction (Stoeglehner et al. 2011c), promoting a mix of functions, density, and nearness to achieve compact cities, towns, and villages. For urban development, the European City (European Commission 2011b), New Urbanism (CNU 2001), EcoCity (Newman and Jennings 2008) can be named, on the regional scale the spatial distribution of functions is envisaged in planning visions such as decentralized concentration (Motzkus 2002), where regional planning shall target a balanced regional distribution of functions, with daily needed functions in the vicinities of the dwelling, and more specialized, not daily required functions concentrated in regional centers. What is normally promoted to guarantee a high quality of life for the population also supports the energy turn.

Yet, the actual spatial development often counteracts against these planning visions and planning principles. In reality, in most countries throughout the world we see low-dense, land-, energy-, and resource-consuming settlement patterns concerning different spatial functions. Mix of functions in walking distance is often decreasing, orienting mobility in the best case—from an energy efficiency point of view—toward public transport, but very often also toward cars as dominating means of transport. When mono-functionality is combined with low density, cars often remain the only meaningful means of transport.

The provision with renewable energy will impose new land use conflicts, which will challenge spatial planning: Speaking in the language of the ecological footprint, fossil fuels are borrowed land from the past (Wackernagel and Rees 1993). They are mined and burned. Carbon, which originates from biologically productive land that lived millions of years ago, is brought to the atmosphere. As mankind intends to use less fossil fuel, it means that we no longer borrow land from the past, but that energy provision generates new land demands in the present. Each of us has a certain resource garden, from which food is provided, industrial raw materials are produced, housing areas and infrastructures are accommodated, and a certain percentage is used for nature protection and for energy land. If the reduced amount of energy from fuels is not spared by energy saving and energy efficiency measures, it is very likely that the share of energy land in the resource garden has to be increased —but which other shares should be reduced? Discussions such as food versus fuel (see, e.g., Hellegers et al. 2008; Catic and Rujnic-Sokele 2008) or the phenomenon of "land grabbing," inter alia, for biofuel production (UN News Centre 2011) are signs that the questions of how to place renewable energy generation in the land use patterns are not reasonably answered yet.

So what we see is that spatial development often goes in the wrong direction, whereas planning visions would support the energy turn. Structures shape behavior, i.e., life styles and economic practices, and behavior shapes structures. Spatial structures also determine the potentials to accommodate renewable energy generation. Therefore, planning for the energy turn has to thoroughly consider spatial

1.2 Spatial Structures and the Energy Turn

structures, as they provide frameworks and affect the possibilities to achieve energy savings and to increase renewable energy provision. As spatial structures are long-lasting, more emphasis has to be laid on their design, as they may open great opportunities or may become a heavy burden for the implementation of the energy turn.

1.3 Scope of the Book

This book addresses the fundamental interrelations between spatial planning and energy planning and shows how these two fields can be integrated in an interdisciplinary approach. We show which fields of action for integrated spatial and energy planning can be identified, and which measures and tools can be implemented on the local and regional level to support the energy turn.

As pointed out above, spatial structures are only one of many framework conditions for the energy turn, but very important ones. We make the interrelations between spatial structures and energy systems visible in order to empower planners, decision makers, and the public in both planning domains to develop consistent strategies: (1) Spatial planners and decision makers shall be able to incorporate energy efficiency and renewable energy supplies in their plans to a much larger extent than is practiced now; (2) energy planners and decision makers shall be enabled to take not only technological, economic, and environmental aspects into consideration, but also to put their strategies and actions in a spatial context, so that potentials and challenges are more realistically determined and a smooth implementation of measures can be supported; and (3) the public shall be empowered to recognize (a) how everyday actions and behavior and substantial decisions such as the choice of a dwelling influence energy demand and supply, and (b) that democratic legitimation and support for the energy turn is crucial to reach climate protection targets and an overall sustainable development.

Our book summarizes the results of eight years of research conducted by the authors of completed and ongoing projects that comprise (1) system analyses, (2) action research integrating energy issues into spatial planning processes, (3) the analyses of legal frameworks, (4) the development of planning tools, (5) integrated spatial and energy concepts for the implementation of energy saving and renewable energy technologies, (6) smart city projects, and (7) the reflection of scientific guidance of a process between the relevant Austrian government levels for the definition of integrated spatial and energy planning. These activities have enabled us to look at the topic from different angles, considering both the factual base for establishing the field and grasping the concerns of a multitude of stakeholder groups affected by the topic.

In this book, we first explain the system interrelations between spatial structures and energy supplies, taking energy efficiency, renewable energy generation as well as energy distribution into account. Second, we lay out which options for the realization of the energy turn are provided by different spatial structures, namely urban areas, suburban areas, small towns as well as rural areas. Third, we provide more detail about the fields of action for integrated spatial and energy planning. Fourth, we introduce measures for integrated spatial and energy planning from the regional to the neighborhood scale. Finally, we present some tools that can support participatory learning processes for integrated spatial and energy planning at the local and regional level.

References

Aleklett, K. (2012). *Peeking at peak oil*. New York: Springer Science & Business Media.
Bailyn, M. (1994). *A survey of thermodynamics*. New York: American Institute of Physics Press.
BFE—Bundesamt für Energie Switzerland (2015). Energiestrategie 2050 der Schweizerischen Eidgenossenschaft. http://www.bfe.admin.ch/themen/00526/00527/index.html?lang=de#. Accessed November 08, 2015.
BMLFUW—Bundesministerium für Land- und Forstwirtschaft, Umwelt und Wasserwirtschaft, BMWFW—Bundesministerium für Wissenschaft, Forschung und Wirtschaft Austria (2010): Energiestrategie Österreich. http://www.bmwfw.gv.at/Ministerium/Staatspreise/Documents/energiestrategie_oesterreich.pdf. Accessed November 08, 2015.
BMWi—Bundesministerium für Wirtschaft und Energie Germany (2015): Die Energiewende gemeinsam zum Erfolg führen. http://www.bmwi.de/BMWi/Redaktion/PDF/Publikationen/die-energiewende-gemeinsam-zum-erfolg-fuehren,property=pdf,bereich=bmwi2012,sprache=de,rwb=true.pdf. Accessed November 08, 2015.
Catic, I., & Rujnic-Sokele, M. (2008). Agriculture products—food for living beings or for machinery. *Gummi, Fasern, Kunststoffe, 61*(11), 701–708.
CNU—Congress for the New Urbanism (2001): Charta of the New Urbanism. https://www.cnu.org/sites/default/files/charter_english.pdf. Accessed November 08, 2015.
Dittmar, H., & Ohland, G. (2004). *The New transit town: best practices in transit-oriented development*. Washington, DC: Island Press.
European Sustainable Energy Innovation Alliance (eseia) (2014): Innovation challenges towards the rational use of bio-resources in Europe—a discourse book. ESEIA. http://www.eseia.eu/files/attachments/10457/453058_eseia_Discourse_Book_May_2014.pdf. Accessed December 17, 2015.
European Commission (2011a): Energy Roadmap 2050. COM(2011) 885 final. http://eur-lex.europa.eu/legal-content/EN/ALL/?uri=CELEX:52011DC0885. Accessed November 08, 2015.
European Commission (2011b): Cities of Tomorrow. Challenges, Visions, Ways forward. http://ec.europa.eu/regional_policy/sources/docgener/studies/pdf/citiesoftomorrow/citiesoftomorrow_final.pdf. Accessed November 08, 2015.
Erker, S., Neugebauer, G., & Stoeglehner, G. (2015). Energieraumplanung—die Energiewende als neue Aufgabe für die Raumplanung. *zoll + Österreichische Schriftenreihe für Landschaft und Freiraum, 1*, 70–73.
Exner, A., Politti, E., Schriefl, E., Erker, S., Stangl, R., Baud, S., et al. (2016). Measuring regional resilience towards fossil fuel supply constraints. Adaptability and vulnerability in socio-ecological transformations—the case of Austria. *Energy Policy, 91*, 128–137.
Farr, D. (2008). *Sustainable Urbanism: urban design with nature*. Hoboken, New Yersey: John Wiley & Sons.
Gaffron, P., Huismans, G., & Skala, F. (2005). *Ecocity book I: A better place to live*. Vienna: Facultas Verlags- und Buchhandels AG.
Gaffron, P., Huismans, G., & Skala, F. (2008). *Ecocity book II: How to make it happen*. Vienna: Facultas Verlags- und Buchhandels AG.

References

Gwehenberger, G., Narodoslawsky, M., Liebmann, B., & Friedl, A. (2007). Ecology of scale versus economy of scale of bioethanol production. *Biofuels, Bioproducts and Biorefinery, 1*(4), 264–269.

Heinze, G. W., & Kill, H. H. (1995). Das Auto von morgen in unseren Städten von morgen. In H. Appel (Ed.), *Stadtauto—Zielkonflikt von Mobilität, Ökologie, Ökonomie und Sicherheit* (pp. 41–69). Wiesbaden: Vieweg.

Hellegers, P., Zilbermann, D., Stedto, P., & McCornick, P. (2008). Interactions between water, energy, food and environment: evolving perspectives and policy issues. *Water Policy, 10*, 1–10.

Lienau, C. (1995). *Die Siedlungen des ländlichen Raumes*. Braunschweig: Westermann Schul-buchverlag GmbH.

Magnago Lampugnani, V. (2010). *Die Stadt im 20. Jahrhundert—Visionen, Entwürfe, Gebautes*. Berlin: Wagenbach.

Motzkus, A.-H. (2002). *Dezentrale Konzentration—Leitbild für eine Region der kurzen Wege?* Auf der Suche nach einer verkehrssparsamen Siedlungsstruktur als Beitrag für eine nachhaltige Gestaltung des Mobilitätsgeschehens in der Metropolregion Rhein-Main. Bonner Geographische Abhandlungen 107. Sankt Augustin: Asgard.

Neugebauer, G., Kretschmer, F., Kollmann, R., Narodoslawsky, M., Ertl, T., & Stoeglehner, G. (2015). Mapping Thermal Energy Resource Potentials from Wastewater Treatment Plants. *Sustainability, 7*(10), 12988–13010.

Newman, P., & Jennings, I. (2008). *Cities as sustainable ecosystems: principles and practices*. Washington, DC: Island Press.

IPCC—Intergovernmental Panel on Climate Change (2014). Climate change 2014: Synthesis report. In Core Writing Team, R. K. Pachauri, & L. A. Meyer (Eds.), *Contribution of working groups I, II and III to the fifth assessment report of the intergovernmental panel on climate change*. Geneva: IPCC.

Prehal, A., & Poppe, H. (2003). *Siedlungsmodelle in Passivhausqualität*. Berichte aus Energie- und Umweltforschung 1/2003. Wien: Bundesministerium für Verkehr, Innovation und Technologie.

Ramirez Camargo, L., Zink, R., Dorner, W., & Stoeglehner, G. (2015). Spatio-temporal modeling of roof-top photovoltaic panels from improved technical potential assessment and electricity peak load offsetting at a municipal scale. *Computers, Environment and Urban Systems, 52*, 58–69.

Regionalmanagement Oberösterreich (2012). Mühlviertler Ressourcenplan, http://www.rmooe.at/projekte/m%C3%BChlviertler-ressourcenplan. Accessed December 17, 2015.

Register, R. (2002). *Ecocites: building cities in balance with nature*. Berkeley California: Berkeley Hills Books.

Schriefl, E., Schubert, U., Skala, F., & Stoeglehner, G. (2009). Urban development for carbon neutral mobility. *World Transport Policy & Practice, 14*(4), 25–35.

Stoeglehner, G., Erker, S., Abart-Heriszt, L., & Neugebauer, G. (2015). *Klima- und Energiemonitoring für die örtliche Raumplanung*. Unpublished project report.

Stoeglehner, G., Erker, S., & Neugebauer, G. (2014a). *Energieraumplanung. Materialienband*. In Zusammenarbeit mit der ÖREK-Partnerschaft "Energieraumplanung". ÖROK Schriftenreihe Nr. 192. Wien: Bundesministerium für Land- und Forstwirtschaft, Umwelt und Wasserwirtschaft, Geschäftsstelle der Österreichischen Raumordnungskonferenz (ÖROK).

Stoeglehner, G., Erker, S., & Neugebauer, G. (2014b). *Tools für Energieraumplanung. Ein Handbuch für deren Auswahl und Anwendung im Planungsprozess*. Wien: Bundesministerium für Land- und Forstwirtschaft, Umwelt und Wasserwirtschaft.

Stoeglehner, G., Narodoslawsky, M., Steinmüller, H., Haselsberger, B., Eder, M., Niemetz, N., et al. (2010). *INKOBA—Durchführbarkeit von nachhaltigen Energiesystemen in INKOBA parks*. Wien: Final report.

Stoeglehner, G., Narodoslawsky, M., Baaske, W., Mitter, H., Weiss, M., Neugebauer, G. C., et al. (2011a). *ELAS—Energetische Langzeitanalysen von Siedlungsstrukturen*. Wien: Final report.

Stoeglehner, G., Narodoslawsky, M., Steinmüller, H., Steininger, K., Weiss, M., Mitter, H., et al. (2011b). *PlanVision—Visionen für eine energieoptimierte Raumplanung*. Wien: Final report.

Stoeglehner, G., Niemetz, N., & Kettl, K.-H. (2011c). Spatial dimensions of sustainable energy systems: new visions for integrated spatial and energy planning. *Energy, Sustainability and Society, 1*(2), 1–9.

UN News Centre (2011). Guidelines to prevent 'land grabbing' crucial for food security, UN expert warns. http://www.un.org/apps/news/story.asp?NewsID=39910#.Vj_XZLxQPJF. Accessed November 08, 2015.

UNEP—United Nations Environment Programme (2002): Melbourne principles for sustainable cities. In *International environmental technology centre, integrative management series nr. 1.* http://www.unep.or.jp/ietc/focus/melbourneprinciples/english.pdf. Accessed November 08, 2015.

Wackernagel, M., & Rees, W. (1993). *How big is our ecological footprint—a handbook for estimating a community's carrying capacity,* Vancouver: University of British Columbia.

WIFO (2011). Energy transition 2012\2020\2050 strategies for the transition to low energy and low emission structures. http://www.wifo.ac.at/publikationen?detail-view=yes&publikation_id=41198. Accessed December 17, 2015.

Chapter 2
System Interrelations Between Spatial Structures, Energy Demand, and Energy Supply

Gernot Stoeglehner, Michael Narodoslawsky, Susanna Erker, and Georg Neugebauer

Abstract Based on a system analysis of elements dealing with spatial structures, energy demand, and energy supply, the most effective regulatory elements for integrated spatial and energy planning are identified. Based on these regulatory elements, the connections between energy efficiency and spatial structures, renewable energy provision and spatial structures, as well as energy logistics and spatial structures, are discussed. Finally, this system analysis is complemented by principles derived from the concept of resilience against energy crises. This approach results in a set of regulatory elements and steering principles to pursue integrated spatial and energy planning.

In order to describe the highly complex system interrelations between spatial structures as well as energy demand and supply, we utilize a system analysis that was carried out in the research project PlanVision (Stoeglehner et al. 2011b). Therefore, systems are understood as explanations of reality, which define the relations of phenomena and influencing factors—hereafter called system elements. The selection of system elements and the identification of their interactions depend on the task of the respective survey. Therefore, systems are always bound to the human perception of complex issues with the following characteristics (Heizinger 1995; Röpke 1977):

- They are "holistic," contrasting the reductionist logic of linear cause–effect relationships; system approaches move from the entity to the details, giving priority to the interrelations between the system elements instead of the detailed descriptions of single elements.
- They are "open," so they are entities, which are in constant exchange with their environments, and their internal interactions are oriented on a common target of the system.
- They are eminent, which describes their property that interactions within the system can result in a new system behavior, which cannot be explained by the features and behavior of the single system elements. In other words, a system is more than the sum of its elements.

In order to derive guidance for action from the system analysis, cybernetics was applied, which deals with the possibilities to steer and regulate complex systems, especially seeks for positive and negative feedback effects (Bertalanffy 1949; Wiener 1948; Vester 2007). Concerning ecosystems and organisms Vester (2007) defines, inter alia, the following main principles of biological cybernetics:

- Negative feedback effects have to dominate over positive feedback, e.g., to prevent uncontrolled growth such as cancer;
- The functioning of the system must be independent from quantitative growth;
- The system must be oriented toward function, not to production;
- Multi-shift use of products, functions, and organizational structures should be applied;
- Cycle processes should guarantee recycling.

Applying system's theory approaches in spatial contexts is an established practice (see, e.g., Vester 1983; Lippuner 2005) The system's theory approach in PlanVision was used to determine the system elements of spatial and energy planning based on the literature surveys, brainstorming, and reflection workshops carried out by an interdisciplinary research team of 19 researchers including disciplines such as landscape, spatial and environmental planning, engineering, energy technology, social sciences, macroeconomics, law, environmental system sciences, and transport science, as well as environmental and natural resource management. Group discussion results were visualized in mind maps (according to Buzan and North 2005; Stoeglehner et al. 2006) and analyzed. The process resulted in a set of 34 system elements stemming from the domains of spatial planning (19 elements) and energy planning (15 elements). In order to operationalize the search for the relevant relations between the system elements, an adapted "paper computer" (Vester 1976, 1980, 2007) was applied. The paper computer is a networking matrix that aims at quantifiable conclusions about the structure of a system and to identify the relevant steering elements of a system by making the impacts of a system element on all other system elements visible. According to their ability to impact other system elements and their exposure to being influenced by others, system elements can be classified into four categories:

- Active elements: These strongly impair other elements, but are only impacted weakly by others.
- Passive elements: These are strongly influenced by other elements and influence others to a small extent.
- Critical elements: These show a strong impact on other elements and are strongly affected by others.
- Buffering elements: These have low effects on others and are weakly affected by other elements.

Concerning the steering of a system, active elements are of high importance, because changing them has a high impact on the system, but the risk of unforeseen rebound effects is low. Starting with the change of critical system elements would also show high impacts, but as they are also influenced by many other elements the chance is high that unforeseen and uncontrolled effects can be detected throughout the system. Trying to change a system with passive or buffering elements is rather hopeless, as impacts on other system elements are weak, and the efforts for change would either be very high or blow out with little effect.

The original model after Vester judges the size of the effect in an ordinal scale and then sums and quotients are calculated. In this way, ordinal data are treated like metric data (Bortz 2005); furthermore, the reliability and inter-subjectivity of the method is not guaranteed, as the scaling of the effects is based on personal judgments and perceptions which are subject to permanent change. In order to overcome this methodological critique of the paper computer, in PlanVision it was determined if a relation between system elements exists (1) or not (0). In a first step, all researches had to judge the interrelations in the matrix. In a second step, all interrelations were discussed in workshops and the interdisciplinary team had to jointly agree on the classifications and reason them. The amount of relations of each system element was counted and then graphically evaluated which element belongs to which category of system elements.

Finally, out of the 34 system elements six turned out to be active, 7 elements to be critical, 13 to be passive, and 8 to be buffering. The system elements under survey and their assessment can be derived from Fig. 2.1. The system elements can be used to analyze policies, legal frameworks or plans if objectives and measures address the active and critical elements, or if they target system elements, which are to be called passive or buffering in a holistic systems analysis. By carrying out such

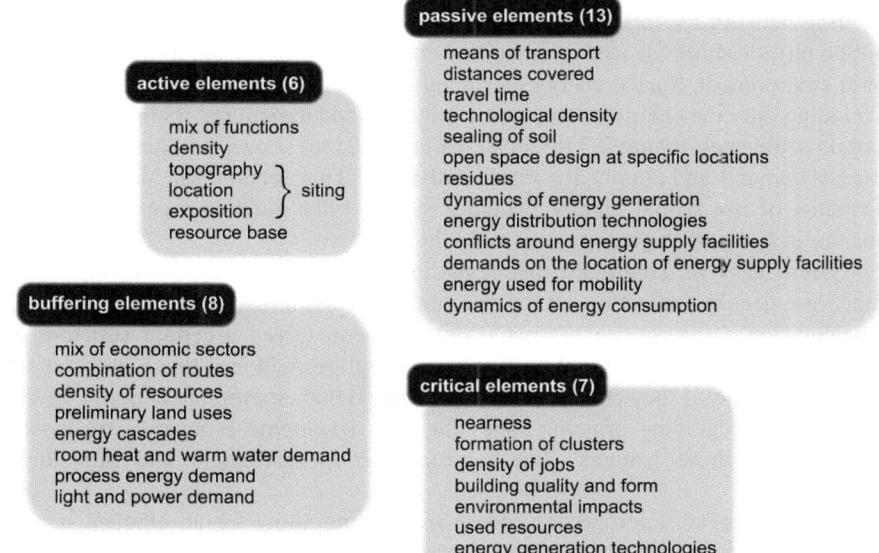

Fig. 2.1 System analysis spatial and energy planning (own illustration after Stoeglehner et al. 2011b)

an analysis, the effectiveness and efficiency of policies, legal frameworks, and plans in supporting the energy turn can be evaluated. The system elements can also be used as criteria to define energy-optimized urban development projects as was already executed in an action research case in an Austrian rural small town, Freistadt (Mandl and Hartl 2011; Stoeglehner et al. 2011b).

Some results concerning the categorization of system elements are rather obvious, whereas other classifications surprised. It has to be considered that the system boundaries are relatively wide and take both spatial and energy planning matters into account. According to the definition of systems laid out before, different system boundaries will very likely lead to different judgments. The analysis of this holistic system shows, inter alia, why some sectorial policies turn out to be very effective, while others fail. Such interpretations can be done because such a crosscutting approach reveals interrelations of phenomena that would be overseen by sectorial approaches. The most important results will be discussed in the next sections.

2.1 Energy Efficiency and Spatial Structures

Our work revealed that besides technological energy efficiency and energy efficiency of lifestyles and economic practices, we can identify a further category of energy efficiency: "spatial energy efficiency" (Stoeglehner et al. 2014b). The concept of "spatial energy efficiency" takes a holistic view on energy demand considering demand for room heating and warm water, mobility, and maintenance of public infrastructures as well as embodied energy in buildings and infrastructures of built environment. Spatial energy efficiency targets the overall energy requirement, excluding only the energy consumed by industry (and agriculture if applicable) in a certain settlement. Figure 2.2 (Stoeglehner et al. 2014a) shows the interrelations for energy demand and supply for residential areas. This section will outline which elements of the spatial energy planning system offer good leverage for system change in favor of the energy turn and why.

System analysis confirmed that regarding energy efficiency a mix of spatial functions, density, and siting are the three most important features for spatial energy efficiency on a macroscale of spatial development. Given a certain structure or energy provision technology, the same supply facilities will lead to a higher overall energy efficiency if they are located in multi-functional, appropriately dense spatial contexts. As a guiding principle, sites for new developments have to be chosen according to these features. This is true both for greenfield and brownfield developments.

Multi-functional, appropriately dense settlements allow for an efficient use of grid-bound energy infrastructure for reasons explained below in detail. Density is a measure of spatial efficiency and includes area values such as population density, density of workplaces, which also lead to increased technical densities, e.g., heat density, which makes energy distribution grid operation more efficient. From a

2.1 Energy Efficiency and Spatial Structures

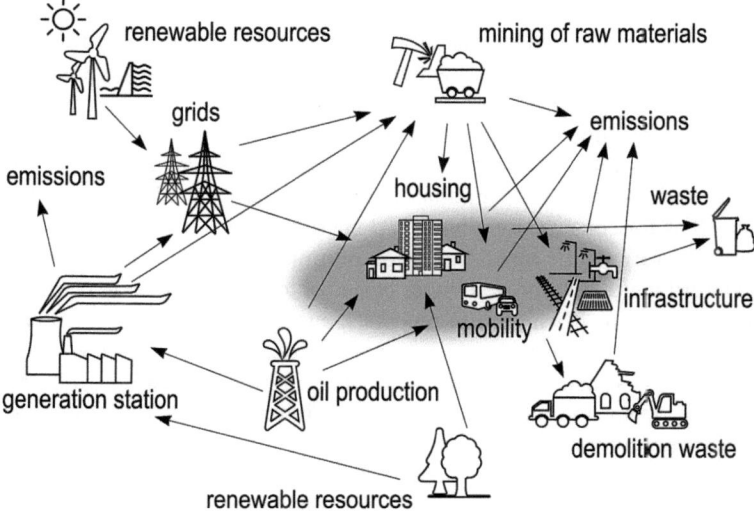

Fig. 2.2 Energy aspects of residential settlements (own illustration after Stoeglehner et al. 2014a)

mere technical-economic point of view, it would be reasonable to postulate that denser spatial structures might be better as long as the building efforts and costs do not exceed the optimum because of too intricate building measures.

This applies not only to construction and maintenance, but density also minimizes the embodied energy in the built structures. Embodied energy is an underestimated factor in spatial development. Evidence from case studies suggests that especially in low-dense residential areas with single-family houses the share of embodied energy in the technical infrastructure (roads, sewer system, water supply) can amount to 55 % of the total energy demand of a settlement, being responsible for up to 90 % of the settlements CO_2 lifecycle emissions. The share of mobility is low, because in low-dense settlements a small amount of persons is living. If state-of-the-art construction techniques for houses are applied, the energy demand for room heat is low. Also the share of house construction is relatively low—normally 5–10 % of the settlement's energy balance when discounted over the life span of buildings. This can make construction and operation of infrastructure the biggest energy consumer in residential areas, which can be easily reduced by denser building schemes. If the same amount of persons is located in three-to-five-story apartment buildings, the share of embodied energy in infrastructure decreases to 5–10 % of the total energy balance, and the share of embodied energy in buildings is likely below 5 % of the total energy balance. If density is higher, the share of energy demand directly related to infrastructure and embodied energy decreases, whereas the demand related to energy services for inhabitants such as mobility and residential heating increases (Stoeglehner et al. 2011a).

Yet, there are limits for density as means to achieve energy efficiency: Settlements can get too dense. Actually, there are two dimensions to these limits, technical and social. In principle, medium-dense spatial structures are more energy efficient than low-dense structures. Technical limits for energy efficiency arise if the energy demand to support buildings reaches a tipping point of too much density and becomes higher again on a per inhabitant basis as for less dense systems. This can happen for extreme high-rise buildings that require considerable energy for operation as well as for construction (embodied energy). Social problems resulting from high density constitute the other dimension of limitation. This may motivate people who can afford it either to move to other places or to look for a second home. For an Austrian small town, it was found out that new medium-dense residential developments, such as row houses, led to a decrease of density in the town as most of the row house dwellers (about 80 %) moved out from high-rise buildings in the same town, fleeing from too cramped living conditions (Emrich Consulting n.y.). Furthermore, people might look for second homes in the countryside. Therefore, infrastructures such as energy, waste, and wastewater treatment in the areas with a high amount of second homes are heavily underused and run on high costs. Furthermore, people traveling from the urban centers to their rural retreats induce increased mobility, mostly by individual car transport.

As a side effect, real estate values decrease in too dense areas, especially when open space is not sufficiently considered in designing an appropriate mix of functions. In Vienna may be observed that in some very dense areas lacking open space the amount of people with low incomes and a migration background is significantly higher than in areas with lower densities and better access to open spaces (Aigner 2013), which might even spur gentrification processes. Therefore, the issue of "economy of scale" versus "ecology of scale" is also valid in the context of density: If the density is too low, proper (energy) infrastructures cannot be provided, so it would make sense to ascribe minimum densities to certain spatial developments. If the density is too high, technical infrastructure and embodied energy may become large. On top of that not only negative effects on the quality of life of the population occur, but the induced behavior to compensate for these effects leads to a higher energy demand of society: (1) More energy is needed and biologically productive land consumed for construction and maintenance of additional buildings and infrastructures in new developments or second homes' areas because of people fleeing from their initial surroundings. (2) Additional mobility is generated for recreational purposes, often with individual cars as the least environmentally friendly means of transport. Finally, not only negative impacts on the energy balances of society and the environment can be detected, but also negative social effects through to gentrification processes. Therefore, "ecology of density" has to be taken into account by assigning minimum and maximum densities for spatial developments. These density values should be considered in building schemes in all kinds of land uses. Even from the energy viewpoint, these density values have to be guided not only by energy efficiency issues, but also by quality of life for the population. Low quality of life might lead to increased energy consumption—if the population possesses enough economic power to take remedial action. These actions may lead to massive rebound effects decreasing the overall density of spatial structures when second homes are included and even higher ecological and economic pressures when increased individual car mobility is factored in.

Mix of functions is an important feature for spatial energy efficiency, as it (1) increases the efficiency of energy grid systems, (2) allows for energy cascades, and (3) has an enormous impact on mobility. First, mix of functions very likely increases the amount of full operating hours of an energy grid system. Each spatial function has a characteristic load function on different timescales. For instance, in winter the room heat demand is highest in the morning and evening and on weekends in residential areas. Complementarily, in commercial areas, the room heat demand is highest during the day and low during nights and weekends. If a grid supplies a single-function area, e.g., a residential area, the variation between base load and maximum load is high, whereas the varied load curves in mixed-use areas will dampen these differences with a high probability leading to a more stable overall load curve which in turn will increase efficiency of the energy provision installation. This is due to the fact that all provision technologies run at considerably lower efficiencies when not operated at full capacity. In Austrian climatic conditions and settlement patterns, district heating grids in pure residential areas normally reach 1500–2200 full load hours per year, whereas mixed-use areas normally reach 4500 full load hours per year. If highly energy intensive facilities such as hospitals, spas, indoor swimming pools are served by the grid, full load hours can increase up to 6000 hours per year, leading to considerable increases in energy provision efficiency (Neugebauer et al. 2015).

Energy systems are especially efficient if the spatial structures allow for energy cascades (see Fig. 2.3). Heat sources such as power plants and industrial facilities provide heat at different temperature levels. In many cases, heat sinks also require heat

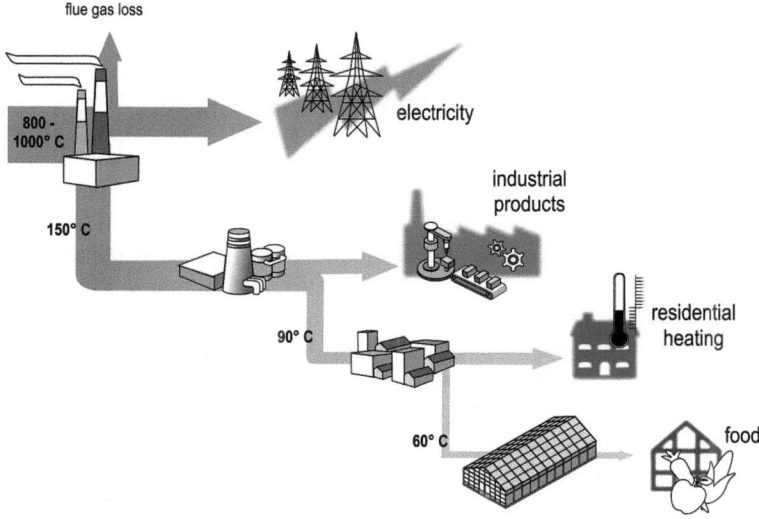

Fig. 2.3 Energy cascade (own illustration)

at different temperature levels. A cascading use of heat now means that from sources to sinks are matched in a way that leads to minimal primary energy input to the whole system. This usually takes the form of using heat that is leaving a certain element in the system to heat another element, so that the heat flow is "cascading" from a high temperature level where it supplies a consumer requiring this high-energy quality to consumers requiring lower heating temperatures (Ayres et al. 1998).

Heat cascades can take very different shapes and depend on the context. An example is a combined heat and power (CHP) plant linked to a district heating system. The heat generated by combustion (depending on the fuel in a temperature range between 800 and 1300 °C) drives a steam turbine, which needs this high temperature to achieve high electricity generation efficiency. The heat of the flue gas leaving the chimney is lost to the environment. The steam leaving the turbine (if it is a back pressure turbine) may be condensed at a temperature level of 150 °C, raising steam to heat an industrial process. Condensing this steam generates heating water at 90 °C that is distributed in a district heating system to residential areas with conventional heating systems. By heating these houses, the water in the grid is cooled to 60 °C, enough to operate low-temperature heating systems in well-insulated state-of-the-art buildings. The water flowing in the grid is further cooled to 40 °C by serving these high-energy standard houses. This temperature may be still high enough to heat greenhouses (Dragone and Rumi 1970) bringing the water temperature in the backflow to the power station to 30 °C (where it is again heated to 90 °C and starts the heating cycle anew).

Heat cascades of course must follow the first law of thermodynamics. It is not the "same" kWh that cascades through all these steps and serves these consumers. The fuel in the CHP plant has to provide the heat for the sum of the demand of all consumers. The only heat loss, however, is the heat lost with the flue gas of the CHP plant. If the CHP plant is replaced by a pure power plant, the heat loss through the flue gas would be comparable, but in addition the heat cooled away by condensing the steam after the turbine is lost as well. It is this heat that serves all the consumers in the cascade and is put to work rather than escaping to the environment. In addition to that, by consciously planning for energy cascades, maximum full load hours considerably exceeding 6000 h can be achieved in grid operation (Lund et al. 2014).

For mobility, a mix of functions is extremely important for environmentally friendly means of transport as well. As structure influences behavior (as well as behavior influences structures) mixed-function, appropriately dense areas spur environmentally friendly means of transport such as walking, biking, and public transport. In such areas, there is no need to use the car, so significantly more daily ways are covered by these means. This can be demonstrated on mobility surveys of municipalities (see, e.g., Stoeglehner et al. 2011a), and mix of functions as a means of reducing travel time and distance is an important claim in many urban planning visions. When we look into greenhouse gas balances of industrialized countries, mobility is a main contributor to climate damaging emissions. In Austria's climate balance emissions from private sectors are decreasing, whereas mobility spoils the balance the most (Umweltbundesamt 2013). Yet, considerable efforts are made in

2.1 Energy Efficiency and Spatial Structures

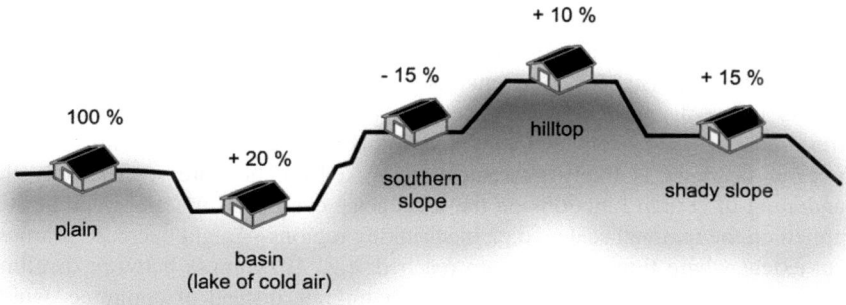

Fig. 2.4 The influence of siting on energy demand (own illustration after EnergieAgentur.NRW 2008; Treberspurg 1999)

the transport sector, but with little effect. Surprisingly, system analysis indicates that mobility is a passive element. Passive elements, however, offer no good leverage for system change. This observation corresponds to the real development, which reveals steady increases of car traffic except for well-organized urban centers and small towns. Simultaneously, spatial development in many regions is dominated by sprawl. This is all the more regrettable as spatial structures are very long lasting.

Furthermore, siting is an important factor for spatial energy efficiency. Not only that siting of new developments—both greenfield and brownfield—decides about the implementation of the mix of functions, but it also takes up the topography and exposition of certain areas. Topography can only be changed on the microscale. On the macroscale, together with exposition it can only be taken into consideration as a boundary condition, which determines, e.g., the amount of direct sunlight and the active and passive use of solar energy. Figure 2.4 shows the influence of siting on the energy demand of residential houses and the relevance for spatial energy efficiency under Central-European climatic conditions. On the microscale, also the design of specific locations has an impact on spatial energy efficiency, but can be influenced rather easily if considered in planning processes: For instance, the shadowing of trees or other buildings can be easily avoided if taken into account in building schemes.

Critical system elements related to spatial structures are nearness, formation of clusters, density of jobs as well as building quality and form. They may also induce system change, but as more rebound effects through more complex cause–effect relationships in the system are to be expected, they have to be treated more carefully. This is especially true when different spatial scales, e.g., regional and local, are taken into consideration. Nearness is on the border between active and critical elements and influences the choice of means of transport, as it has an effect on travel times and distances. Problematic is the subjective character of nearness, as it is highly influenced by individual perception and depending on certain regional and local spatial contexts. Therefore, nearness should be assessed related to travel time budgets instead of distances (Cerwenka et al. 2000). A specifically critical aspect of nearness is the distance between dwelling, work, and daily supply, which also corresponds to the system elements formation of clusters and density of jobs. However, nearness is likely guaranteed in mixed-function areas, especially the abandoning of certain functions on the local scale even if all functions can be still supplied on the regional scale—e.g., in shrinking regions—might decrease nearness to an extent where tipping points are reached: e.g., If nearness between dwelling and work decreases too much, people might migrate instead of commute, which leads to a destabilization of the local population, an increase in second homes and to accelerated spatial development in labor market centers with the respective increase of energy demand concerning embodied energy for additional building efforts.

Clusters are economic agglomerations around one or few thematic areas such as products and/or services, e.g., automotive industry clusters or IT industry clusters. The actors of clusters, such as companies, research institutes, universities, are closely linked with each other (Springer Gabler Verlag n.y.). Clusters have specific influence on accessibility and the energy demand for (industrial) processes. As long as the products and services are demanded on the markets, clusters might provide high economic prosperity for certain regions, but if demand falls, this might cause severe economic crises in the affected regions and a high loss of jobs carving negative development spirals. In contrast, a mix of economic sectors on a regional scale is a buffering system element, stabilizing the system and providing for more opportunities to cope with crises. These interrelations are relevant for the energy systems as they have influence on mobility patterns and their respective energy demand, the energy demand of production processes as well as on the resource base for the regional energy provision, especially concerning energy cascades.

On the microscale, the building quality and form are critical system elements, as they heavily influence the range of energy technologies that can be applied in a certain spatial context. Concerning already-existing built structures, a high-energy demand is due to low energy efficiency of buildings. Taking single-family houses in Austria into account, at a current refurbishment rate of about 1 % of the building stock per year, it takes something around 100 years to adapt existing buildings to contemporary buildings standards (Baumgartner et al. 2010). In newly developed areas, demand from low energy houses can be minimal, and in combination with on-site renewable energy provision technologies, even an energy surplus can be reached. Such developments very likely make, for instance, district heating grids in newly developed or restored areas obsolete. Therefore, the feasibility of grid-bound energy systems has to be regularly re-evaluated. If tipping points by the broad application of energy-efficient building technologies are reached, system change from centralized to decentralized energy provision is likely to happen. As such change has severe economic and environmental consequences, it has to be permanently monitored and taken into consideration while planning visions and measures.

2.2 Renewable Energy Provision and Spatial Structures

To shift from a fossil and nuclear energy provision to a renewable resource base means to increase land demand for energy supplies (see the resource garden metaphor in Chap. 1). Therefore, additional land uses or multiple use of land (e.g., when solar panels are mounted on roofs or wind turbines are erected on pastures) are introduced into already intensively used areas, causing considerable land use change and making land use conflicts very likely. Therefore, an important task for integrated spatial and energy planning will be to search for agreeable locations for renewable energy installations on the one hand and to protect present and future resource provision areas on the other hand (Stoeglehner et al. 2014b).

Existing spatial structures heavily influence the possibilities to introduce energy provision facilities, as can be illustrated by wind energy: Wind energy impacts, compared to other energy provision technologies, relatively few environmental issues negatively. Except on bird and bat protection and landscape sceneries, the impacts can be resolved by safety distance approaches, e.g., protection from noise and shadowing (Felber and Stoeglehner 2014). Therefore, in order to build wind parks with minimal conflict, spatial structures with huge areas between settlements are needed. Given the fast development of technologies with spin wheels becoming bigger and bigger, even higher distances between wind parks and settlements are needed. This leads to the conclusion that sprawl situations complicate or prevent wind energy use, whereas mixed-function, appropriately dense areas save open space and offer more possibilities for the use of renewable energy sources.

Although our system analysis covers 14 system elements concerning energy supplies, only one system element is evaluated as being active and three as being critical. The active system element is the resource base, which determines the critical system elements technologies and related environmental effects. From a planning point of view, it is therefore more promising to plan the resource mix in the energy supply in planning visions than focusing too early on technologies, although these issues cannot be completely separated. The focus on resources can also help mitigate environmental effects by taking ecological capacity limits in renewable energy provision into account. For instance, the amount of biomass production can be determined on the local and regional scale considering the regional ecosystem limits. Which maximum rates of provision of certain energy sources, may this be biomass, wind or solar energy integrated in buildings, are environmentally feasible and socially accepted can be expressed based on regional energy potential surveys. The determination of capacity limits can take the societal value base into account, expressed, inter alia in the food versus fuel debate, the toleration of landscape change, etc. When the resource base for a renewable energy supply, which can be environmentally feasibly extracted, is defined in a certain spatial context, technology portfolios to make best use of this resource base can be agreed on. In the choice of technologies, socioeconomic issues can be taken into consideration.

On the microscale locating energy generation facilities, their environmental impacts as well as their resources have the character of critical system elements.

The selection of sites as well as the type and size of the plant, which are determined by the resources applied, accounts for possible conflicts about land uses as well as social acceptance. If not enough environmentally and socially feasible sites are available for a certain determined technology option, this poses a considerable threat for master-planned future energy systems that were mainly based on technological and economic analyses. Spatial analysis will make such energy plans even more feasible by considering regional and local conditions, yet it is possible that stakeholder involvement, or even the resistance of the public against a certain element, e.g., a certain energy generation plant, might delay or prevent a desired energy planning option by the respective decision-makers, threatening the applicability of the whole energy action plan. In order to resolve this problem, we dedicate Chap. 6 to planning processes and tools.

2.3 Energy Logistics and Spatial Structures

Energy logistics is a key problem concerning the technological networks supporting the energy turn, as renewable energy sources have very different characteristics than fossil and nuclear energy sources that have to be replaced. These different characteristics of renewable energy sources strongly relate to the spatial dimensions of the energy turn.

Solar radiation does not provide the same energy density than fossil resources, as they are typically decentral and bound to area in contrast to fossil and nuclear resources that are typically point resources, emerging from a mine or bore hole. Figure 2.5 shows the energy-harvesting density for some renewable energy sources compared to fossil crude oil. This is the energy that may be harvested from one m^2 per year.

Figure 2.5 shows that the energy harvest density in kWh/m^2 year is high for solar heat, even higher than that for crude oil (measured as the annual energy output from a Middle East oil field divided by its area). All other renewable energy forms have distinctly lower energy harvest densities than crude oil. While the harvesting density for solar heat is high, transport of heat is limited as will be discussed below.

Fig. 2.5 Energy-harvesting density for different energy resources (Narodoslawsky 2014)

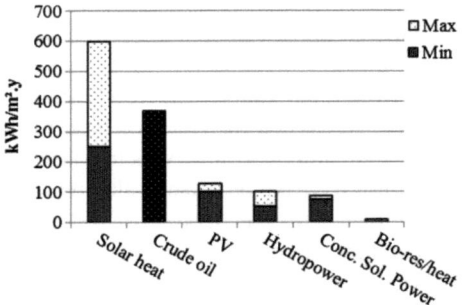

2.3 Energy Logistics and Spatial Structures

This makes solar heat an interesting option for settlements to convert solar radiation directly to heat, using the area of roofs. It will compete, however, with PV which shows a significantly lower harvesting density but provides electricity, an energy form of much higher quality. Both PV and hydropower, although with much lower energy-harvesting density than crude oil, may still warrant their utilization in central installations. The energy-harvesting density for biomass is, however, dramatically lower than that of other sources. This is due to the low efficiency of transformation of solar radiation to usable energy of plants. This means that the logistics of collecting biomass is a major challenge and that decentral energy conversion technologies based on biomass have distinctive logistical advantages compared to large centralized units (eseia 2014).

Digging a little deeper into the spatial connotation of the use of bioresources reveals the magnitude of this challenge. Table 2.1 shows humidity, density, and energy density (based on incineration for relatively dry materials and biogas production for wet materials).

This table reveals that energy densities for bioresources differ by an order of magnitude and are in any case considerably lower than those of fossil materials. The logistical challenge becomes even more visible if different means of transportation are factored in. According to their transport efficiency (and strongly influenced by their particular ratio of empty weight to load capacity), different means of transportation require different energy to transport a load over a certain distance. If the limit of the energy used to transport a resource to its utilization site is arbitrarily set to 1 % of the contained energy, the following results are obtained (Narodoslawsky 2014):

Table 2.1 Transport parameters for different bioresources compared to fossil energy carriers (own table after Gwehenberger and Narodoslawsky 2008)

Conversion	Material	Humidity (%w/w)	Energy content (MJ/kg)[a]	Density (kg/m^3)[a]	Energy density (MJ/m^3)[a]
Incineration	Straw (gray)	15	15	100–135	1500–2025
	Wheat (grains)	15	15	670–750	10,050–11,250
	Rape seed	9	24.6	700	17,220
	Wood chips	40	10.4	235	2440
	Split logs (beech)	20	14.7	400–450	5880–6615
	Wood pellets	6	14.4	660	9500
Biogas production	Grass silage	60–70	3.7	600–700	2220–2590
	Corn silage	65–72	4.2	770	3230
	Organic municipal waste	70	2.4	750	1800
	Manure	95	0.7	1000	700
	Light fuel oil	0	42.7	840	36,000
	Anthracite	0	35.3	800–930	28,000–33,000

[a]All numbers are related to fresh material

- In the case of manure, straw, and corn silage, 1 % of the contained energy will power a tractor (as the most common short distance means of transportation on farms) 5, 7, 12, or 18 km, respectively;
- 1 % of the energy contained in wood chips and split logs will power a truck for 40 and 100 km, respectively;
- For wood pellets and corn, a train will go for 475 and 525 km, respectively, using 1 % of the transported energy content;
- An ocean going ship loaded with crude oil, however, will travel 7800 km with 1 % of the energy contained in its cargo.

Low-grade bioresources and biowastes offer major potential as sources for sustainable energy systems as they do not compete with other functions of bioresources such as nutrition or raw materials for timber, pulp, or paper industry. These numbers mean that these bioresources may, however, only be utilized close to the points of their emergence. Making low-grade bioresources transportable over longer distances requires decentralized treatment facilities. Treatment can include different methods: The simplest is drying, such as wood chips to be burned in combined heat and power (CHP) cycles for electricity generation and district heating, or district heating alone. Also sludge from wastewater treatment has, if dried to a water content of 20–30 %, e.g., by solar–thermal energy or by the use of waste heat, a calorific value in a comparable order of magnitude as brown coal (DWA 2012; Böhmer et al. 2001). If dried sludge is processed in mono-incineration plants and the ashes are separately stored, they can be used for phosphor recovery (ÖWAV 2014), so that energy and resource generation can be combined.

When it comes to biomass, the concept of biorefineries is increasingly used to exploit as much resource potential as possible: First, biomass can be transferred to biofuels or platform chemicals, i.e., semi-products that can be used as raw materials for chemical processes, allowing for a non-fossil resource base of the chemical industry (Stoeglehner et al. 2010). Wastes and low grade by-products of biorefineries may then be utilized to generate power and heat. Both the off-heat from process energy of the biorefinery as well as the heat from CHP-based electricity generation can be provided by district heating grids. Generally, CHP systems work more technically and economically efficient the higher the demand for heat is. This is heavily influenced by the spatial structures that can be provided with a district heating system, as already stated above. On the one hand, multi-functional, appropriately dense spatial structures should be aimed for supply areas. On the other hand, a second conclusion about the planning of CHP supply is suggested (eseia 2014): The size of a CHP system should be oriented on the heat demand, so that the available heat can be used by customers, making electricity the by-product. CHP systems trying to maximize electricity output without looking at the usability of the heat side are less energy efficient and have a higher risk to fail economically. In Austria can be observed that due to high biomass prices, electricity-only biogas plants cannot be run with an economic surplus at the moment (E-Control 2014). CHP plants hint at still another spatial aspect of bioenergy technologies: They link different distribution grids. In more general terms, intersections of distribution grids are particularly favorable sites for either biorefineries or bioenergy technologies based on lower grade bioresources. Many bio-based energy technologies provide energy in different forms. CHP plants, regardless if based on incineration or biogas, in general provide electricity and heat. Biogas fermenters may supply CHP plants and upgrading units as well. These upgrading units separate methane that may be injected into the gas grid. The same holds true for synthetic natural gas (SNG) plants that gasify solid bioresources and may again supply CHP plants and methane injection. All these technologies "hybridize" energy systems and may switch between different distribution grids according to the demands for different energy forms.

A second challenge arises from the fact that some renewable energy sources such as solar energy and wind energy show cyclical and intermittent time dependencies in their supply. As PV and wind power are dependent on natural phenomena, and energy demand, however, is following economic and behavioral rules, discrepancies between demand and supply of energy are inevitable. In addition, the electrical grid as the densest distribution grid offers very limited storage capacity, especially for storage of energy over time spans exceeding minutes. Spatiotemporal modeling is an appropriate tool to determine within a certain spatial context, e.g., a settlement, a town: (1) how supply and demand curves correlate; (2) which amount of energy in which time spans can be directly used without delivery to a grid or has to be either stored or requires grid capacity to be transported to other places of demand; or vice versa, (3) during which time periods electricity has to be taken from the grid to cover the demand. A spatiotemporal analysis of photovoltaic (PV) use in a residential settlement under Middle-European climatic conditions and consumer patterns revealed that about 20–40 % of the yearly energy demand can be accommodated in the energy system without a high storage or transportation demand, whereas PV provision levels above 40 % of the yearly energy demand make either storage or grid capacity necessary (Ramirez et al. 2015).

Besides trying to overlap demand and supply by using spatiotemporal analysis, there are other options to overcome the intermittent nature of important renewable energy sources that have profound impacts on spatial structures. Storage is of course an obvious candidate. It is, however, a fact that storage costs are high for electricity, the energy form whose distribution grid has the lowest storage capacity.

Figure 2.6 compares different electricity storage options and shows that using batteries for storage of excess electricity is prohibitively expensive. The figure indicates that a hybridization of grids and the link between grids may hold considerable potential for making energy distribution both smart and stable. Integrating bioresource-derived electricity to stabilize grids is certainly one option, but here decentral CHP plants must cooperate within a smart grid architecture due to the logistical parameters of bioresources discussed above. As mentioned already, the size of CHP plants should be defined by the heat demand, and its operation, however, must follow the demand of the electricity grid. This then leads to the necessity of storing heat as its generation is technically coupled to electricity generation. Heat, however, is easily and cheaply stored, at costs of less than 10 % of the cheapest way to store electricity (Narodoslawsky 2014).

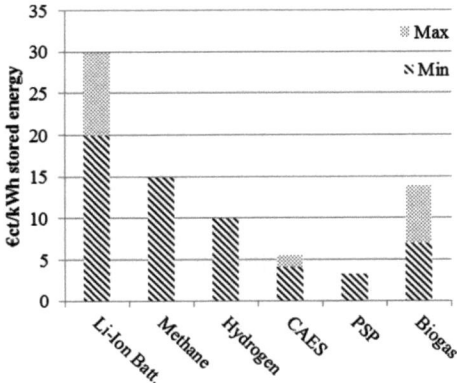

Fig. 2.6 Cost for electricity storage (Narodoslawsky 2014). Methane refers to methane generated by power-to-gas processes, Hydrogen produced by electrolysis, CAES Compressed Air Energy Storage, PSP Pump Storage of Power, Biogas produced from grass silage

Other challenges arise from using excess electricity generated by wind power and PV to produce hydrogen and methane in power-to-gas systems. This calls again for linking the gas and electricity grids and, in the case of methane production, the optimization of siting of conversion plants as they need carbon dioxide from either fossil- or bioresource-based power plants as well as other industries as cement production (Reiter and Lindorfer 2015).

A third major challenge is linked to the different properties of energy distribution systems. They differ widely in their storage capacity and the losses when transporting energy over a certain distance. Assuming the same 1 % of energy loss for transporting (after Narodoslawsky 2014), this will transport the following:

- natural gas via a gas grid at full capacity for 250 km,
- electricity in a high-voltage (380 kV) grid over 100 km and
- in a medium voltage grid (110 kV) over 17 km, and
- heat in a district heating system for less than 1 km.

It is obvious that these numbers again have strong spatial implications. Both high-voltage electricity and gas grids are clearly interregional, even continental in their distribution characteristic. Heat by contrast is distinctly local. Following the argument pursued before, any CHP installations must be sized to serve local heat demand and may use interregional grids to distribute higher value energy or support their stability. Conversely, as a challenge to spatial planning, heat surplus emerging from utilizing regional low-grade bioresources should meet spatial structures that are able to absorb this heat.

Regarding storage capacity, heat and gas grids show high capacity for storing intermittently provided energy. The electricity grid, however, is least tolerant regarding intermittent energy provision. This means that operation of grid-overarching energy provision technologies must be oriented to the needs of the electricity grid while storing energy preferably in the form of heat and gas.

The challenges discussed above are general and apply to rural as well as urban and metropolitan spaces, however, with different emphases. Smart city development in particular must, among other factors, include these considerations about the spatial connotations of low-carbon energy systems. The arguments above, however, reveal that the planning for smart cities does not stop at the city perimeters: resources as well as distribution systems for energy systems require a holistic spatial view that also includes the city hinterland as it has to consider resources and distributions systems in the settlement area itself.

2.4 Energy Resilience and Spatial Structures

In light of upcoming challenges regarding climate change and the energy turn common strategies, concepts, and models concerning spatial planning are getting supplemented with new approaches and terminologies. This also applies to the new concept of resilience and the attempt of developing more resilient spatial structures.

Generally, resilience is a much-discussed concept and is used differently by various disciplines. From an overall perspective, resilience can be described as characteristics of social, environmental, or economic systems comprising the ability to (1) preserve the core functions, structures, and the identity of a system in case of a shock, stress, or disturbance, (2) stabilize disturbed processes after the event within a given period of time, or (3) enable the system to adapt and reorganize (Bourbeau 2013; Birkmann 2008; Holling 1973, 2001; Walker et al. 2004). Therefore, both the duration and the spatial dimensions of the disorder play a decisive role and more or less determine the successful way of handling a crisis (Carpenter et al. 2001).

In order to cope with disturbances, resilience-based planning does not need to focus on missing but on existing system characteristics, such as regional resources, social skills, and adaptive capacity (O'Brien 2009). Furthermore, resilience must be understood as a continuous process not a desired target state, which means an ongoing development toward an often unknown direction. Long periods without crises or stress might lead to routine and may imply a wrong sense of security, stability, or balance. Resilient planning anticipates and accepts changing circumstances and, therefore, attempts to deal with uncertainties in advance (McAslan 2010; Bohle 2007). Only by considering these perpetual processes of change, potential crises can be detected ex ante and the respective system might be prepared for no longer unexpected and convertible challenges. Therefore, resilience thinking includes learning processes about the adaptation of the value base, visions, objectives, and action strategies in light of observed and perceived changes. In the context of integrated spatial and energy planning, resilience, thus, calls for a periodic monitoring and evaluation of spatial development and energy systems and the respective inter-linkages. By providing measures concerning the cognition of future risks, communities are able to preliminarily assess and adjust their visions and have the opportunity to choose between alternative planning options. Therefore, the integration of resilience thinking into current planning system can help to recognize upcoming problems at an early stage in order to consequently identify potential solutions.

Analyzing the current energy system with the concept of resilience in mind, the predominant demand and supply of fossil fuels is conspicuous. Fossils are concentrated in relatively few geographic areas and have to be transported over long distances, which results in a high level of dependence on these energy sources. Due to this focus, the resilience of the energy system to a possible energy crisis is low, which makes the system vulnerable. Herein, vulnerability refers to the exposure, the sensitivity, and the susceptibility to suffer of damage in the case of certain external disturbances (Gallopin 2006). As opposite of vulnerability, the ability to preserve systemic structures and properties during disturbances can be mentioned (Gallopin 2006), which herein is defined as congruent with the first level of resilience (see, definition of resilience).

As energy crises could be a possible trigger for stress and disturbance within energy systems, they have to be further discussed. Basically, an energy crisis can be seen as either a restriction of access to one or several energy source(s) or a physical limitation of an energy source, e.g., by exploitation reaching the capacity limit of a source (see, e.g., Schabbach and Wesselak 2012). Consequently, energy crises can be caused by, e.g., a rise of energy prices, global uncertainties, conflicts between energy exporting and importing states, technical malfunctions, attacks on energy-related infrastructures, or by a decreasing availability of an energy source. Therefore, energy crises concerning fossil fuels such as oil or natural gas would have far-reaching global consequences, such as effects on the nutrition, the mobility sector, or the heat supply. At the same time, spatial structures are getting affected by changes in the energy system and influence the resilience of the energy system.

To give an example, the spatial developments of the past two centuries were characterized by a change in energy sources, especially the transition toward fossils, and the development of new technologies especially in the mobility sector, such as cars. Consequently, compact industrial cities could evolve into sprawling, service-, consumer-, and car-oriented settlements due to the utilization of oil and gas. By implementing renewable energy sources into our energy system, significant adjustments of spatial structures could be expected (Fischer 2014; Sieverts 2012).

We argue that resilience needs to be deliberately implemented in processes of integrated spatial and energy planning in order to improve planning strategies, leading us to the concept of "spatial energy resilience," which we introduce here. This concept expresses the possibilities to increase resilience of energy systems by means of spatial planning. According to the previous definitions, we distinguish six dimensions of spatial energy resilience as depicted in Fig. 2.7 (after Godschalk 2002; Birkmann and Fleischhauer 2009; Beatley 2009; Greiving et al. 2009).

Herein, the principle "ability to learn" builds the overall framework for resilient planning and affects the successful implementation of other principles. Learning is a process and means to acquire new, modified or existing knowledge, behaviors, (social) skills, and abilities (Weber 1977; Wilkesmann 1999). Thus, it is a key to cope with new circumstances (Bower and Hilgard 1983). However, in the case of an energy crisis, it will be necessary not only to learn from past events but also to

2.4 Energy Resilience and Spatial Structures

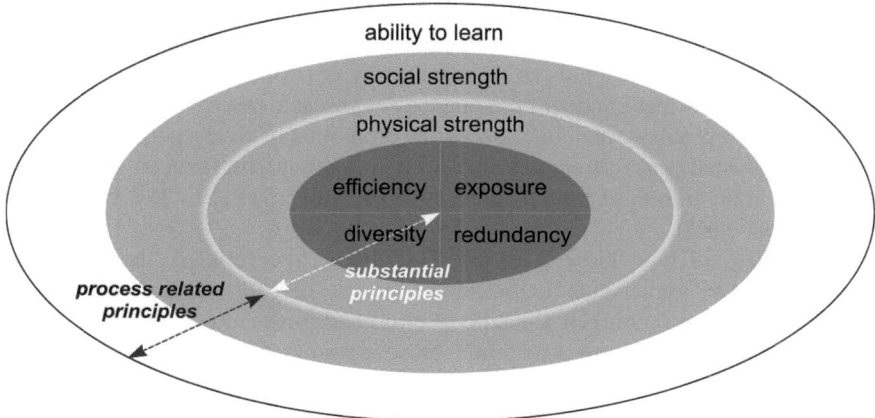

Fig. 2.7 Principles of spatial energy resilience (own illustration)

anticipate future challenges in order to scrutinize and adapt paradigms and own opinions. Therefore, the ability to learn in all its dimensions is fundamental in order to increase resilience. Integrated spatial and energy planning is able to support this principle by creating and preparing fundamental information for decision-making in order to reinforce knowledge, to sharpen the focus of planning debate, and to consider different solutions.

Following this principle, "strength" constitutes a further framework condition and constitutes the link between the overall framework and the remaining structural principles. The overarching aim of this principle is to increase robustness and resistance of existing and planned structures and, thus, to reduce the negative consequences of crises. Strength thereby means the ability of a system to compensate or prevent dysfunction (Norris et al. 2008). In the context of spatial energy resilience, strength has a physical and a social dimension. Herein, the physical dimension refers to the substantial principles, whereas the social aspect of strength is related to the process-oriented principle "ability to learn."

Concerning the physical dimension, characteristics, which lead to spatial energy efficiency as well as to better spatial preconditions for renewable energy supplies and energy logistics, also improve the resilience against energy crises. Therefore, the considerations of the preceding subchapters support physical strength and have to be implemented in accordance with the central principles.

In addition, the social dimension of strength can be identified as properties which increase cohesion and cooperation. This refers to the concept of "social capital," which describes the relationships between individuals, social groups, organizations, or different hierarchical levels (Bourdieu 1986; Coleman 1988; Putnam 1993). Therefore, it is important to support and balance three kinds of relationships: bonding, bridging, and linking. In this context, bonding refers to the like-mindedness of homogeneous individuals and groups. Bridging characterizes the openness to other social structures and values and, therefore, tends to connect people across various social groups. Finally, the term linking expresses the vertical relations between different hierarchical levels (Coleman 1988; OECD 2002; Woolcock and Sweetser 2002). Social capital can be activated by providing facilities where people get the chance to meet and to communicate with each other. In this regard, the revitalization and supply of public spaces, the promotion of civic engagement as well as a strengthened sense of community through a strong club structure need to be supported in order to raise social capital and cope with crises (OECD 2004; Putnam 2001; Bohle 2005).

Depending on how these framework conditions are fulfilled, the substantial principles can be implemented: efficiency, exposure, diversity, and redundancy. As "efficiency" means optimizing the ratio of effort and result, in the context of spatial energy resilience it refers to possibilities to save resources. This complies with the concept of energy-efficient spatial structures as introduced in Sect. 2.1. Despite this positive effect of efficiency, it can also be interpreted as economic strategy to reduce effort to a minimum, which means to provide each function only once in the cheapest possible way—which also would mean at the lowest possible, but still sufficient quality. Therefore, it is discussed if efficiency is a concurring concept to resilience (Löhr 2009). We argue that efficiency interpreted as means to save natural resources is part of resilience, but cannot be separated from the other issues such as diversity and redundancy. Efficiency as part of resilience means that diverse and redundant subsystems are present, but that each of them is also efficient in order to save as much resources as possible and, therefore, also to reduce the ecological footprint of spatial structures and energy systems to a minimum.

This principle is followed by "exposure," which means to reduce the spatial dispersion of settlement structures and infrastructure as well as to minimize the dependencies on energy sources. Self-sufficiency, a reduction of resource and energy consumption and the focus on regional resources in order to guarantee autonomy and independency, can be summarized under this principle.

"Diversity" stands for the promotion of a mixture of different spatial functions and a wide range of energy sources at different locations. The different realizations and unequal design of systems or components provide a broad range of sensitivity to a given disorder. Therefore, some components or systemic parts might stop working under shock while others remain functioning. Diversity promotes the flexibility of a system and gives the opportunity to switch between heterogeneous components. This might lead to a variety of multiple approaches, possible solutions, and opens up room for maneuver.

Finally, the set of principles of resilience in the context of integrated spatial and energy planning is getting completed by "redundancy." In order to support this principle, a number of functionally same or similar components must be created. This increases the systemic security in the event of a crisis. In the case of system failure, the additionally installed component can allow proper functioning. In addition, redundant systems should be installed at different locations in order to minimize the risk of a common vulnerability. In the context of resilient spatial and energy planning, redundancy means to implement structures as well as to enable multiple uses of structures. Herein, this principle must be seen as complementary rather than competing with efficiency, exposure, and diversity. Although similarly functioning components, such as power generation and power distribution systems, can be designed in an efficient way, their intended function has to be covered several times. Decentralized energy systems might be of special value for redundancy, since they tend to increase the resilience by replacing large power plants with a multitude of small energy producers (Stelter 2009). Therefore, the balance of the substantial principles is essential and has to be adequately implemented according to the particular spatial context, as will be pointed out in Chap. 3.

By considering these six principles of resilience, the persistence, adaptability, and transformability of systems might be ensured and, therefore, the quality of planning processes can be improved. The concept of spatial energy resilience must be seen in addition to the considerations of the former subchapters, as they focus on structural and contextual aspects of integrated spatial and energy planning. Spatial energy resilience includes process-related components and connects them with the content level. Therefore, the active system elements should be brought in line with the objectives of resilience. The spatial- and energy-related elements "mix of functions," "density," "siting," and "resource base" can operationalize the substantial principles of resilience: "physical strength" including "efficiency," "redundancy," "diversity," and "exposure." Implementing these system elements could be supported by the process-oriented resilience principles "social strength" and "ability to learn."

References

Aigner, J. (2013). *Freiräume in Ottakring - Gleiche Chancen für alle? Analyse des öffentlichen Freiraumangebotes des Bezirks Ottakring anhand eines neuen Wiener Modells zur Grün-und Freiraumversorgung*. Master Thesis, University of Natural Resources and Life Sciences Vienna (supervisors: Gernot Stoeglehner, Franz Grossauer).

Ayres, R. U., Ayres, L. W., & Martins, K. (1998). Exergy, waste accounting, and life-cycle analysis. *Energy, 23*(5), 355–363.

Baumgartner, G., Gutschi, G., Bachhiesl, U., & Lackner, A. (2010). Technisch-wirtschaftliche Analyse energetischer und thermischer Sanierungsmöglichkeiten von Einfamilienhäusern. 11. Symposium Energieinnovation, 10.-12.2.2010, Graz/Austria. https://online.tugraz.at/tug_online/voe_main2.getvolltext?pCurrPk=49779. Accessed 08 Nov 2015.

Beatley, T. (2009). *Planning for coastal resilience*: best practice for calamitous times. Washington, DC: Island Press.

Bertalanffy, L. (1949). *Allgemeine Systemtheorie*, Biologica Generalis.

Birkmann, J. (2008). Globaler Umweltwandel, Naturgefahren, Vulnerabilität und Katastrophenresilienz. Notwendigkeit der Perspektivenerweiterung in der Raumplanung. *Raumforschung und Raumordnung, 66*(1), 5–22.

Birkmann, J. & Fleischhauer, M. (2009). Adaptation strategies for spatial development to climate change: "Climate proofing"—outline a new planning tool. *Raumforschung und Raumordnung, 67*(2), 114–127.

Bohle, H.-G. (2005). Social or unsocial capital? The concept of social capital in geographical vulnerability research. *Geographische Zeitschrift, 93*(2), 65–81.

Bohle, H.-G. (2007). Geographische Entwicklungsforschung. In: Gebhardt, H., Glaser, R., Radtke, U., & Reuber, P. (Eds.), Geographie. Physische Geographie und Humangeographie (pp. 797–815). Heidelberg: Elsevier/Spektrum Akademischer Verlag.
Böhmer, S., Rumplmayer, A., Rapp, K., & Baumgartner, A. (2001). Mitverbrennung von Klärschlämmen in kalorischen Kraftwerken. Umweltbundesamt GmbH (Ed.). UBA-BE-194, Vienna.
Bortz, J. (2005). *Statistik für Human- und Sozialwissenschaftler.* Heidelberg: Springer Medizin Verlag.
Bourbeau, P. (2013). Resiliencism. Premises and promises in securitisation research. *Resilience, 1*(1), 3–17.
Bourdieu, P. (1986). The forms of capital. In J. G. Richardson (Ed.), *Handbook of theory and research for the sociology of education* (pp. 241–258). New York: Greenwood.
Bower, G. H., & Hilgard, E. R. (1983). *Theorien des Lernens.* Stuttgart: Klett-Cotta.
Buzan, T., & North, V. (2005). *Mind Mapping. Der Weg zu Ihrem persönlichen Erfolg.* Wien: öbv & hpt Verlag.
Carpenter, S., Walker, B., Anderies, J. M., & Abel, N. (2001). From metaphor to measurement: Resilience of what to What? *Ecosystems, 4,* 765–781.
Cerwenka, P., Hauger, G., Hörl, B., & Klamer, M. (2000). *Kompendium der Verkehrssystemplanung.* Wien: Österreichischer Kunst- und Kulturverlag.
Coleman, J. S. (1988). Social capital in the creation of human capital. *American Journal of Sociology, 94,* 95–120.
Dragone, G., & Rumi, O. (1970). Pilot greenhouse for the utilization of low-temperature waters. *Geothermics, 2*(1), 918–920.
DWA – Deutsche Vereinigung für Wasserwirtschaft, Abwasser und Abfall e.V. (2012). *Thermische Behandlung von Klärschlämmen – Mitverbrennung in Kraftwerken.* Merkblatt DWA-M 387, Hennef.
E-Control (2014). Ökostrombericht 2014. http://www.e-control.at/publikationen/oeko-energie-und-energie-effizienz/berichte/oekostrombericht. Accessed 17 Dec 2015.
Emrich Consulting (n.y.). Migration behaviour within settlements of different density. Oral communication.
EnergieAgentur.NRW (2008). *Planungsleitfaden. 50 Solarsiedlungen in Nordrhein-Westfalen.* o. V. Düsseldorf.
eseia (2014). *Innovation challenges towards the rational use of bio-resources in Europe—a discourse book.* eseia. http://www.eseia.eu/files/attachments/10457/453058_eseia_Discourse_Book_May_2014.pdf. Accessed 17 Dec 2015.
Felber, G., & Stoeglehner, G. (2014). Onshore wind energy use in spatial planning—a proposal for resolving conflicts with a dynamic safety distance approach. *Energy, Sustainability and Society, 4*(22), 1–9.
Fischer, E. P. (2014). *Unzerstörbar. Die Energie und ihre Geschichte.* Berlin, Heidelberg: Springer.
Gallopin, G. C. (2006). Linkages between vulnerability, resilience, and adaptive capacity. *Journal of Global Environmental Change, 16*(3), 293–303.
Godschalk, D. R. (2002). *Urban hazard mitigation: Creating resilient cities. Plenary paper presented at the Urban Hazards Forum.* New York: City University of New York.
Greiving, S., Fleischhauer, M. & Dosch, F. (2009). Klimagerechte Stadtentwicklung: Rolle der bestehenden städtebaulichen Leitbilder und Instrumente. Bundesamt für Bauwesen und Raumordnung. *BBSR-Online-Publikation, 24.*
Gwehenberger, G., & Narodoslawsky, M. (2008). Sustainable processes—The challenge of the 21st century for chemical engineering. *Process Safety and Environmental Protection, 86*(5), 321–327.
Heizinger, P. (1995). Systemintegrierter Umweltschutz und Saubere Technik – Ein komplexes Phänomen. Dissertation, Technical University Graz.
Holling, C. S. (1973). Resilience and stability of ecological systems. *Annual Review of Ecology and Systematics, 4*(1), 1–23.

References

Holling, C. S. (2001). Understanding the complexity of economic, ecological, and social systems. *Ecosystems, 4*(5), 390–405.
Lippuner, R. (2005). *Raum, Systeme, Praktiken: Zum Verhältnis von Alltag, Wissenschaft und Geographie*, Sozialgeographische Bibliothek Bd.2. Stuttgart: Franz Steiner Verlag.
Löhr, D. (2009). Resilience vs. efficiency—a critical view on the contemporary environmental economic science. *Umweltwissenschaften und Schadstoff-Forschung, 21*(4), 393–406.
Lund, H., Werner, S., Wiltshire, R., Svendsen, S., Thorsen, J. E., Hvelplund, F., & Vad Mathiesen, B. (2014). 4th Generation District Heating (4GDH) Integrating smart thermal grids into future sustainable energy systems. *Energy, 68*(2014), 1–11.
Mandl, M., & Hartl, E. (2011). Generationenwohnen am Bauhofareal Freistadt. https://www.land-oberoesterreich.gv.at/Mediendateien/Formulare/DokumenteAbt_U/Freis-tadt_Generationenwohnen.pdf. Accessed 08 Nov 2015.
McAslan, A. (2010). *The concept of resilience. Understanding its origins, meaning and utility*. Adelaide: Torrens Resilience Institute.
Narodoslawsky, M. (2014). Utilising bio-resources—a rational strategy for a sustainable bio-economy, ita-manuscript, Vienna. http://epub.oeaw.ac.at/ita/ita-manuscript/ita_14_02.pdf. Accessed 08 Nov 2015.
Neugebauer, G., Kretschmer, F., Kollmann, R., Narodoslawsky, M., Ertl, T., & Stoeglehner, G. (2015). Mapping thermal energy resource potentials from wastewater treatment plants. *Sustainability, 7*(10), 12988–13010.
Newman, P., & Jennings, I. (2008). *Cities as sustainable ecosystems: Principles and practices*. Washington DC: Island Press.
Norris, F. H., Stevens, S. P., Pfefferbaum, B., Wyche, K. F., & Pfefferbaum, R. L. (2008). Community resilience as a metaphor, theory, set of capacities, and strategy for disaster readiness. *American Journal of Community Psychology, 41*(1–2), 127–150.
O'Brien, G. (2009). Vulnerability and resilience in the European energy system. *Energy & Environment, 20*(3), 399–410.
OECD. (2002). *Measurement of social capital: The Canadian experience*. London.
OECD. (2004). *Vom Wohlergehen der Nationen*. Paris: OECD Publishing.
ÖWAV – Österreichischer Wasser- und Abfallwirtschaftsverband (2014). *Klärschlamm als Ressource*. ÖWAV-Positionspapier, Wien.
Putnam, R. D. (1993). *Making democracy work: Civic traditions in modern Italy*. Princeton: Princeton University Press.
Putnam, R. D. (2001). *Gesellschaft und Gemeinsinn: Sozialkapital im internationalen Vergleich*. Gütersloh: Verlag Bertelsmann Stiftung.
Ramirez C. L., Zink, R., Dorner, W., & Stoeglehner, G. (2015). Spatio-temporal modeling of roof-top photovoltaic panels from improved technical potential assessment and electricity peak load offsetting at a municipal scale. *Computers, Environment and Urban Systems, 52*, 58–69.
Reiter, G., & Lindorfer, J. (2015). Evaluating carbon dioxide sources for power-to-gas applications —a case study for Austria. *Journal of CO_2 Utilization, 10*, 40–49.
Röpke, J. (1977). *Die Strategie der Innovation – eine systemorientierte Untersuchung der Interaktion von Individuum, Organisation und Markt im Neuerungsprozess*. Tübingen: J.C.B Mohr (Paul Siebeck).
Schabbach, T., & Wesselak, V. (2012). *Energie – Die Zukunft wird erneuerbar*. Berlin Heidelberg: Springer.
Sieverts, T. (2012). Resilienz – Zur Neuorientierung des Planens und Bauens. *disP – The Planning Review, 48*(1), 83–88.
Springer Gabler Verlag (ed.) (n.y.). Gabler Wirtschaftslexikon, Stichwort: Cluster. http://wirtschaftslexikon.gabler.de/Archiv/5140/cluster-v14.html. Accessed 08 Nov 2015.
Stelter, A. (2009). *Siedlungsentwicklung und Energielogistik in Deutschland im Spannungsfeld von Zentralität und Dezentralität*. Frankfurt, M; Berlin; Bern; Bruxelles; New York; Oxford; Wien: Lang.
Stoeglehner, G., Baaske, W., Mitter, H., Niemetz, N., Kettl, K. H., Weiss, M., et al. (2014a). Sustainability appraisal of residential energy demand and supply—a life cycle approach

including heating, electricity, embodied energy and mobility. *Energy, Sustainability and Society, 4*(24), 1–13.

Stoeglehner, G., Erker, S., & Neugebauer, G. (2014b). *Energieraumplanung. Materialienband.* In Zusammenarbeit mit der ÖREK-Partnerschaft „Energieraumplanung". ÖROK Schriftenreihe Nr. 192. Wien: Bundesministerium für Land- und Forstwirtschaft, Umwelt und Wasserwirtschaft, Geschäftsstelle der Österreichischen Raumordnungskonferenz (ÖROK).

Stoeglehner, G., Erker, S., & Neugebauer, G. (2014c). *Tools für Energieraumplanung. Ein Handbuch für deren Auswahl und Anwendung im Planungsprozess.* Bundesministerium für Land- und Forstwirtschaft, Umwelt und Wasserwirtschaft (Ed.). Wien.

Stoeglehner G., Mitter H., & Jungmeier P. (2006). Adult education as a key factor of sustainable rural development. In: Subai C., Ferrer-Balas D., Mulder K.F., & Moszkowicz P. (Eds.), *Engineering education in sustainable development*, 4.-6.10.2006, Lyon; ISBN: 978-2-905015-63-1.

Stoeglehner, G., Narodoslawsky, M., Baaske, W., Mitter, H., Weiss, M., Neugebauer G.C., Niemetz, N., Kettl, K.-H., Eder, M., Sandor, N., & Lancaster, B. (2011a). *ELAS – Energetische Langzeitanalysen von Siedlungsstrukturen.* Final report. Wien.

Stoeglehner, G., Narodoslawsky, M., Steinmüller, H., Haselsberger, B., Eder, M., Niemetz, N., Kettl, K.H., Sandor, N., Kollmann, A., Lindorfer, J., Tichler, R., Fazeni, K. (2010). *INKOBA – Durchführbarkeit von nachhaltigen Energiesystemen in INKOBA Parks.* Final report. Wien.

Stoeglehner, G., Narodoslawsky, M., Steinmüller, H., Steininger, K., Weiss, M., Mitter, H., Neugebauer G.C., Weber, G., Niemetz, N., Kettl, K.-H., Eder, M., Sandor, N., Pflüglmayer, B., Markl, B., Kollmann, A., Friedl, C., Lindorfer, J., Luger, M., & Kulmer, V. (2011b). *PlanVision – Visionen für eine energieoptimierte Raumplanung.* Final report. Wien.

Treberspurg, M. (1999). *Neues Bauen mit der Sonne. Ansätze zu einer klimagerechten Ar-chitektur.* 2. Aufl., Wien: Springer.

Umweltbundesamt (2013). Zehnter Umweltkontrollbericht. http://www.umweltbundesamt.at/umweltsituation/umweltkontrollbericht/ukb/. Accessed 08 Nov 2015.

Vester, F. (1976). *Ballungsgebiete in der Krise: vom Verstehen und Planen menschlicher Lebensräume.* München: Deutscher Taschenbuchverlag.

Vester, F. (1980). *Neuland des Denkens.* Stuttgart: Deutsche Verlags-Anstalt GmbH.

Vester, F. (1983). *Ballungsgebiete in der Krise - Vom Verstehen und Planen menschlicher Lebensräume.* München: dtv.

Vester, F. (2007). *Die Kunst vernetzt zu denken.* 6. Aufl. München: dtv.

Walker, B. H., Holling, C. S., Carpenter, S., & Kinzig, A. (2004). Resilience, adaptability, and transformability in social-ecological systems. *Ecology and Society, 9*(2), 5.

Weber, E. (1977). *Pädagogik. Eine Einführung.* Donauwörth.

Wiener, N. (1948). *Cybernetics or control and communication in the animal and the machine.* Paris: Hermann Editions.

Wilkesmann, U. (1999). *Lernen in Organisationen.: Die Inszenierung von kollektiven Lernprozessen.* Frankfurt, New York: Campus Verlag.

Woolcock, M., & Sweetser, A. T. (2002). Bright ideas: Social capital—the bonds that connect. *ADB Review, 34*(2), 26–27.

Chapter 3
Spatial Archetypes in the Energy Turn

Gernot Stoeglehner, Michael Narodoslawsky, Susanna Erker, and Georg Neugebauer

Abstract Spatial structures and spatial planning decisions have considerable influence on the realization of the energy turn as they shape energy demand on the one hand and potentials of renewable resource-based energy provision on the other hand. Spatial planning has the task of setting frameworks for the energy turn by securing and shaping energy-optimized spatial structures. Taking into account inherent characteristics of renewable resources and the structure of distribution pathways, different spatial archetypes, particularly urban areas, rural areas, rural small towns, suburban areas as well as suburban small towns, provide different options for the realization of the energy turn. These opportunities are depicted in terms of generic visions for the respective spatial archetypes and discussed along the main features of spatial structures, namely mix of functions, density, siting and resources as basis for renewable energy generation and provision. Subsequently, the contributions of the respective spatial archetypes to fulfill the structural objectives of resilience against energy crises are summarized.

Different spatial structures offer different options and emphases to implement sustainable energy systems based on energy efficiency and renewable resources. Therefore, new definitions of energy systems-related functions for different spatial structures have to be considered. This requirement arises from the inherent characteristics of renewable resources (see Sect. 2.3). Compared to fossil resources, renewables are normally characterized by lower yields per hectare (see Fig. 2.5) and, therefore, predestined for decentralized provision. Moreover, bioresources show disadvantageous logistic properties including low transport density and high humidity (see Table 2.1). In addition, the structure of distribution pathways is an important determinant as several energy qualities are distributed via large-scale infrastructures and show widely different ranges and distribution densities.

For the realization of the energy turn, several archetypes of spatial structures can offer particular implementation opportunities. In principle, three kinds of boundaries can be considered for the delineation of such spatial archetypes according to the described methods on the establishment of city or urban area boundaries in the literature: (1) *administrative* boundaries referring to territorial or political boundaries (Hartshorne 1933; Aguilar and Ward 2003), (2) *functional* boundaries taking into account connections or interactions between areas (Brown and Holmes 1971; Douglass 2000; Hidle et al. 2009), and (3) *morphological* boundaries based on the form or structure, e.g., of land use or land cover (Benediktsson et al. 2003; Rashed et al. 2003).

Taking functional connections and interactions into account, the following typology of spatial archetypes can be derived based on the results of the research project PlanVision in order to move toward energy-optimized spatial structures (after Stoeglehner et al. 2011): *Urban areas* represent the main consumers of resources and energy, but also provide complex (industrial) goods (e.g., electronic devices, machinery, cars, etc.) and services. *Rural areas* are the main providers of resources and energy. *Rural small towns* accommodate the resource conversion function and represent an attractive living space for the decentralized industrial society. *Suburban areas* constitute the spatial reserve for urban areas and take over a major supply function for them. *Suburban small towns* can be considered as subdominant centers of the wider conurbation without or with resource conversion only at a low level due to lower availability of arable land. From the viewpoint of integrated spatial and energy planning, the assignment of functions to spatial archetypes is essentially determined by requirements of energy services, energy efficiency, energy sources, and technologies.

On the level of material products, a differentiation can be made between the following types of products: (1) *fresh products* of daily consumption (especially high-quality food products) that allow only short transport distances, (2) *commodities* meaning easy-to-transport standardized bulk goods as a basis for further processing, and (3) *complex* (e.g., industrial) *goods* that are manufactured from commodities with large inputs of knowledge and investments. As depicted in Table 3.1, the provision function for the described spatial archetypes can be determined along this rough classification of products.

Based on current research findings (Stoeglehner et al. 2011; Exner et al. 2016), a classification of the Austrian territory according to the above-mentioned five spatial archetypes can be carried out on the level of municipalities combining territorial and functional criteria. The categorization is based on the criteria (a) population figure (on municipal as well as settlement level), (b) location in urban regions ("Stadtregionen") according to Statistics Austria (2012), (c) membership in the Association of Austrian Cities and Towns (2015), and (d) balance of commuter flows. In a first step, the municipalities are classified according to population figures on municipal level (Statistics Austria 2016) into five size categories. In a second step, this information is overlaid with an assignment of the municipalities to the zonation of the Austrian urban regions "Stadtregionen" according to Statistics Austria (2012). The model for the delineation of urban regions results in a classification comprising two zones (Wonka and Laburda 2010): (1) a core zone with high population density and (2) an outer zone with a high percentage of commuters into the core zone. In contrast to previous delineations that used the localities as spatial reference basis, the current method is based on population and employment

Table 3.1 Matrix of provision and consumption among spatial archetypes (own table after Stoeglehner et al. 2011)

Product type	Consumer	Provider
Fresh products of daily consumption	Urban area	Suburban area
	Rural area	Rural area
	Rural small town	Rural area
	Suburban area	Suburban area
	Suburban small town	Suburban area
Commodities	All archetypes	Rural small town
Bioresources for commodities	Rural small town	Rural area
Complex industrial goods	All archetypes	Urban area
		Suburban area
		Suburban small town
		Rural small town

figures in 500 m grids that were subsequently extended to the municipal level. In a third step, the membership in the Association of Austrian Cities and Towns was considered in order to take the self-perception of the municipalities into account. Finally, based on the balance of commuter flows, the function of a municipality as regional center in terms of attractive working locations was considered in order to identify rural and suburban small towns.

The results of the classification of the 2102 Austrian municipalities are depicted in Fig. 3.1. Austrian urban areas include the federal capital Vienna, four large cities (capitals of federal provinces: Graz, Linz, Innsbruck, and Salzburg), and several medium cities (inter alia three further capitals of federal provinces: Klagenfurt, St. Pölten, and Dornbirn). With a share of approximately 63 % of the Austrian municipalities, the majority can be considered as rural areas. Slightly more than one percent of the Austrian municipalities can be categorized as rural small towns. Approximately, one-third of the municipalities shape suburban areas primarily located around the federal capital and the above-mentioned capitals of the provinces. Some suburban areas are located around one suburban small town or are formed around polycentric structures of several suburban small towns without direct connection to an urban area. Finally, more or less one and a half percent of the Austrian municipalities can be assigned to the category of suburban small towns.

In the following subsections, generic visions for the spatial archetypes such as urban areas, rural areas, rural small towns, suburban areas, and suburban small towns are drafted. The visions define their basic functions and describe them in more detail along the dominating system elements of integrated spatial and energy planning: *mix of functions, density, siting and zoning,* and *resource base* (Stoeglehner et al. 2011). Furthermore, the contribution of the respective spatial archetypes to address the content-related principles of resilience in terms of

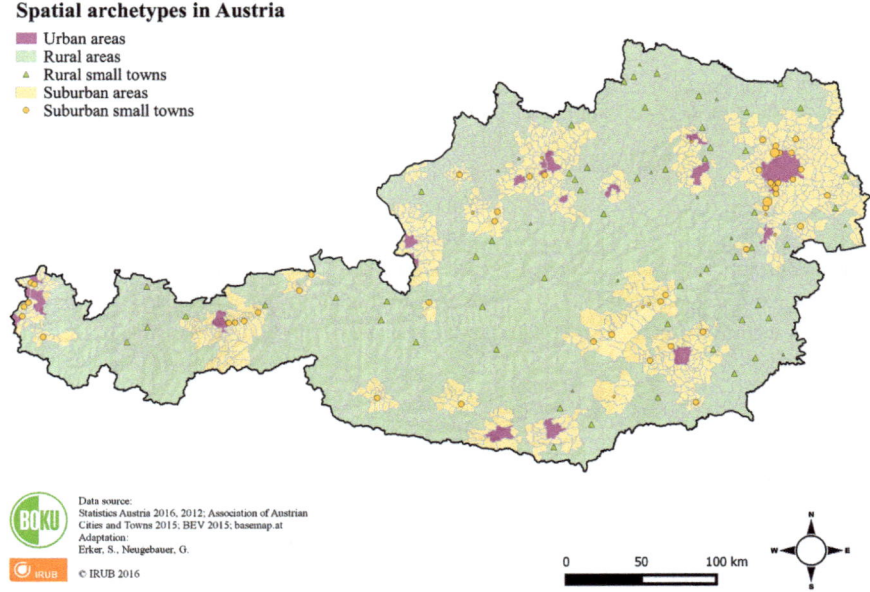

Fig. 3.1 Classification of the Austrian territory according to five spatial archetypes (own illustration after Stoeglehner et al. 2011; Exner et al. 2016)

physical strength (efficiency, exposure, diversity, and redundancy) as set out in Sect. 2.4 is summarized, whereas the process-related principles *ability to learn* and *social strength* are discussed in Chap. 6.

3.1 Urban Areas

With a share of 54 % in 2014, more than half of the global population lives in urban areas (UN DESA 2015). According to the Fifth Assessment Report of the Intergovernmental Panel on Climate Change, urban areas are responsible for about 67–76 % of the global energy use and account for about three quarters of the energy-related CO_2-emissions (Seto et al. 2014; Creutzig et al. 2015). This can be explained, on the one hand, by their size and high population densities and, on the other hand, because urban areas are main locations for the production of industrial and consumer goods and the provision of services. At the same time, a high share of sealed soil and potential land use conflicts set restrictions on renewable energy provision potentials. Accordingly, the main task for sustainable urban development is to increase energy and resource efficiency in order to minimize resource and energy input. Table 3.2 gives an overview of a generic vision for urban areas from a comprehensive energy and resource planning viewpoint.

Table 3.2 Generic vision of energy-optimized spatial planning for urban areas (own table after Stoeglehner et al. 2011)

Basic function	Goal
Living space for majority of people/attractive working location	Achieving and securing highest quality of live
	Providing sufficient leisure time opportunities
	Ensuring high environmental quality
	Guaranteeing comprehensive provision of goods and services, education (up to tertiary level), social (healthcare) and cultural services, research and development
	Providing sufficient and adequate employment
Main energy and resource consumer	Achieving highest utilization efficiency
	Reducing ecological pressures from energy provision/utilization to a minimum
	Minimizing resource consumption
Provider of complex (industrial) goods and services	Achieving highest resource conversion efficiency
	Assuring strong societal interaction
	Ensuring international interconnectedness

With regard to *mix of functions*, in urban areas, the basic residential function is closely linked to educational and research institutions, leisure time opportunities, shopping and gastronomical offers, the service sector, and low- or zero-emission production companies. This provides for multi-functional centers and districts, where the basic spatial functions including recreation are arranged within walking and/or biking distance to the residential function or at least in a distance that can be bridged by means of public transport. Mix of functions can be achieved within building blocks or floor by floor within buildings.

The connection to industrial and commercial areas not suitable for a mixture with the residential function is provided by high-quality, efficient public transport. Provision with goods and the disposal function comprise distribution and transport tasks rather than production issues. Primary production contributes a share to renewable energy generation and to food production.

In terms of technical efficiency, multi-functionality is an important precondition to balance the dynamics of energy demand and generation. During the course of a day, multi-functional areas show an increasing convergence with lower peak demands than single-used areas as load curves of several functions balance out. As energy generation and distribution have to be designed for peak demands, mix of functions might avoid an inefficient operation outside of peak load periods.

Moderate population and employment *densities* are an important precondition for the efficiency of complex infrastructures (e.g., energy supply, public transport, high-quality social infrastructures) and the potential to gain economic advantages due to agglomeration effects. Moderate densities reduce area consumption and consequently ecological pressures and allow for saving space for primary production, recreational purposes, and ecological compensation areas. However, density shall not exceed certain limits to secure quality of life.

Siting and zoning of different land uses decide about the implementation of mix of functions and play an important role in minimizing the energy demand and the local energy generation potential. On a system level, the decision for locating favoured functions in urban areas is of greater importance than the decision between several locations within the urban area. In addition, siting serves for the minimization of land use conflicts and safeguards recreational and ecological compensation areas.

Concerning the *resource base*, the possibilities for renewable energy provision in urban areas are limited due to a high share of sealed soil and potential land use conflicts. Energy provision based on urban resources can be mainly build on thermal use of non-recyclable waste and the exploitation of solar energy for thermal purposes and photovoltaics. In order to achieve highest utilization efficiency, energy provision is organized into energy cascades (see Fig. 2.2) from energy uses with higher exergy to those with lower exergy (Stremke et al. 2011). As potentials for renewable energy sources are limited, energy has to be imported into urban areas from the respective "energy hinterlands" to a large extent.

In the context of the above-discussed resilience concept (see Sect. 2.4), urban areas have to face challenges in strengthening and developing their basic functions in particular regarding the structural principle "*exposure*" due to dependencies on their respective "energy and resource hinterlands." Developing and offering living space for the majority of people as well as the provision of complex (industrial) goods and services are associated with energy and resource consumption on a high level. Energy and resource demands beyond own generation capacities determine the essential vulnerability of urban areas concerning energy and resource provision. Therefore, increasing energy and resource efficiency contributes to minimize exposure by reducing the energy and resource input.

However, urban areas especially feature considerable advantages concerning the further three structural resilience principles if mix of functions and appropriate densities as important features for spatial energy efficiency are implemented. Interweaving basic spatial functions in multi-functional and appropriately densely populated areas enables resource and energy "*efficiency*", e.g., concerning the operation of complex infrastructures. Shaping efficient urban energy systems means an optimization of energy and resource flows, e.g., by exploiting waste heat of industrial facilities, wastewater treatment, or waste incineration plants that can contribute to satisfy heat demands in the vicinity of the facilities via efficient district heating systems.

The integration of several land uses in terms of mix of functions allows for an effective implementation of the "*diversity*" principle. On the one hand, the arrangement of several basic functions in close proximity enables their local accessibility by walking and/or biking so that mobility can be ensured independently from motorized private transport. In combination with adequate public transport systems, also disadvantaged groups (e.g., children, youth, elderly people, or low-income individuals) have access to mobility. On the other hand, the nearness of several spatial functions to each other enables cascading energy use. Multi-functional areas allow for matching heat sources and sinks in an optimal way so that primary energy inputs can be reduced to a minimum. Besides, urban areas have to face also challenges in the context of *"diversity."* Concerning potential sources of energy, urban areas cannot build on the whole spectrum, as some energy generation technologies are not available or not in such an amount as in other spatial archetypes due to potential land use conflicts.

Fulfilling the principle of *"redundancy"* means the implementation of functionally identical or comparable components in order to ensure the systemic security in a crisis situation with a larger capacity than needed in normal operation. In terms of decentral energy supply, a huge amount of small energy generation facilities connected with smart grids replaces large dominating power plants. In particular, solar energy and waste heat constitute the respective potentials of urban areas. Smart grids do not only distribute energy from (partially decentralized) providers to consumers, but offer also information exchange for an efficient adaptation of consumption to generation (and vice versa). This harmonization is of particular importance in order to minimize cost-intensive and inefficient energy storage (see Sect. 2.3). Furthermore, multi-functional areas that offer a diverse local supply represent further advantages of urban areas, and finally, a public transport system that provides different routing options and comprises several mobility forms and mobility providers can contribute to redundancy in covering mobility demands.

3.2 Rural Areas

The future of energy generation and resource production for the whole society lies in rural areas, as they have to ensure the provision of the other spatial archetypes with biomass-based resources and renewable energy that cannot be generated in the other spatial archetypes. Furthermore, rural areas take over the recreation function and offer the residential function and basic supply in balanced relations to the former mentioned basic spatial functions, renewable resource production, energy provision, and recreation. Table 3.3 gives an overview of a generic vision for rural areas from a comprehensive energy and resource planning viewpoint.

Mix of functions in rural areas means interlinking the following functions: As supply area for all other spatial archetypes, rural areas have the task of providing resources for the whole society and guarantee basic supply with goods for daily consumption, childcare, education on primary level, etc. Furthermore, rural areas appear as recreation areas and have responsibility for the long-term securing of biological productivity and stable ecosystems. In particular, this means mix of functions in primary production within environmental capacity limits including also reintroducing of materials and nutrients from conversion processes and from harvest. In addition, mix of functions between several production sections and the respective disposal functions (e.g., biogas incl. manure recirculation) serves the objective of highest logistic efficiency. Concerning settlement development, this means providing living space for the population needed to maintain primary production, basic supply, and recreational infrastructure.

Table 3.3 Generic visions of energy-optimized spatial planning for rural areas (own table after Stoeglehner et al. 2011)

Basic function	Goal
Sufficient population density for primary production and maintenance of basic services	Balanced population with basic provision of goods (daily consumption), education (primary level), social and cultural services
Recreational space	Achieving highest environmental quality
	Providing sufficient infrastructure to ensure the recreation function
Sustainable resource provision	Achieving maximum efficiency in space utilization
	Securing long-term area productivity within environmental capacities
	Ensuring stable ecosystems
	Guaranteeing highest logistic efficiency for renewable resources and by-products of conversion processes
	Avoiding resource import

The dwelling function in rural areas is oriented toward the basic function of providing resources in the long term taking the objectives of securing maximum land productivity and ensuring stable ecosystems into account. *Density* for settlements is important in order to save bioproductive land and ecological compensation areas. Density is of particular importance for the local provision of efficient supply and infrastructure systems, especially with regard to the objective of highest logistic efficiency.

Location choices for residential and supply functions as well as recreational infrastructure are subordinate to the objectives of maximum area productivity and logistic efficiency. Therefore, *siting and zoning* are restricted on existing settlement cores, and a gradual expansion of bioproductive land as well as area for achieving ecological stability (e.g., by decommissioning of unused or heavily underused settlement areas) is desirable.

Rural areas substantially assume the provision function of raw materials for food manufacturing, energy generation, and industrial production although further processing is often not arranged on the production location in rural areas, as the respective enterprises realistically search for the benefits of "economies of scale." In the long term, if the energy turn shall be successful, imports of primary raw materials as well as fossil resources into rural areas are not permitted.

Considering spatial energy resilience, rural areas are of great importance for sustainable development as provider of bioproductive areas. These allow for basic supply as well as energy and resource provision in the long term and represent the basis for physical strength of rural areas in terms of spatial resilience. Concerning "*efficiency*," rural areas show advantages due to the availability of land that justifies focussing on the primary sector with resource and food provision. In addition, energy generation is of great importance. In contrast to urban areas, where walking and biking or efficient public transport systems ensure mobility not reliant on petroleum, in rural areas with spacious structures and large distances between several basic spatial functions, mobility can be based on alternative energy sources. Electric mobility is one option to increase efficiency as the degree of efficiency of an electrically driven vehicle exceeds the fourfold of a petrol or diesel-run vehicle (Pötscher et al. 2010). The composition of the primary energy sources for the power generation plays an essential role. Necessary structural transformations are likely to be understood as chances and benefits for rural areas, as they are also new sources of income for the rural population.

Concerning the structural principle *"exposure,"* on the one hand, rural areas have to face challenges because of dependencies on urban areas and the basic spatial functions located there. In a crises situation, difficult accessibility of agglomerations and their functions may have challenging implications for rural areas. On the other hand, resource production and energy generation largely take place in rural areas so that the short-distance access is guaranteed and the higher the energy generation is, the lower is the exposure to energy crises.

Disposing over the biggest pool of available land, rural areas can make the largest contribution to diversified energy generation. According to local available potentials, energy generation can build on wind energy, solar energy, hydropower, biomass, geothermal, or tidal energy. Due to the *"diversity"* of potential energy sources, bottlenecks of one source can be counteracted with the help of others. A challenge considering diversity of energy generation technologies might arise from additional land use conflicts when energy provision, resource provision, food production, and the residential function enter into competition with each other.

Rural areas dispose over bioproductive land as reservoir that can be used in terms of *"redundancy"* to guarantee the provision function via functionally comparable structures. For instance, decentralized energy provision builds on diverse energy sources. Yet, the low density makes it especially challenging to implement redundant systems from an economic viewpoint.

3.3 Rural Small Towns

In a renewable resource economy, rural small towns will become of increasing importance in the long term, which can be argued by the nature of biomass-based resources. Bioresources are characterized by low durability and low transport density and have to be converted into commodities with higher durability and transport density. Physical proximity from the harvesting area to the sites of the transformation into commodities ensures high transport efficiency, both of fresh products from the fields to the processing plants and of residues from the processing plants back to the fields. The second aspect is especially necessary to close nutrient cycles. Rural small towns show favorable conditions as platforms of resource processing: Waste heat from these transformation processes can be used efficiently in district heating systems in compact settlement structures. Rural small towns are

Table 3.4 Generic visions of energy-optimized spatial planning for rural small towns (own table after Stoeglehner et al. 2011)

Basic function	Goal
Attractive living space for decentralized industrial society	Achieving and securing high quality of life
	Providing excellent leisure time opportunities
	Ensuring high environmental quality
	Guaranteeing advanced provision of goods and services that satisfy upmarket requirements, education (up to secondary level), social and cultural services, research and development
Resource conversion	Reducing ecological pressures from resource conversion/utilization to a minimum
	Gaining highest resource conversion efficiency
	Linking the distribution grids

often located conveniently and have rail connection, so that the further transportation to urban areas can be performed efficiently. Following this, vision might generate new jobs in rural small towns, potentially, in connection with education as well as research and development functions. Table 3.4 depicts a generic vision for rural small towns and their role in the realization of the energy turn.

Rural small towns can be modeled as platforms of resource processing for commodities, which can be justified with the nature of renewable resources (little durability, low transport density). *Mix of functions* means a node function of grids (e.g., information, electricity, transport, district heating). Rural small towns dispose over job and supply functions for the regional population (contrary to suburban areas that are oriented toward urban areas concerning these aspects). Additionally, innovative capabilities can emerge through research and development in the field of commodities provision. This corresponds to the objectives to shape small towns as attractive living space for decentralized industrial society on the one hand and resource conversion on the other hand (especially highest resource conversion efficiency and optimal management of supply grids). Rural small towns have sufficient potential to utilize residual materials and energy and, furthermore, meet the supraregional demands for commodities. Concerning the linkage of basic spatial functions, mix of functions in rural small towns complies with the objective of high quality of life via accessibility in walking and/or biking distance.

Settlement *density* in rural small towns contributes to establish environmentally friendly transport (e.g., bus, train and bike) on the regional scale. Due to nearness, transport in rural small towns often comprises walking and biking, whereas public transport is mainly important in regional and supraregional contexts. Creating density is one major aspect in linking distribution grids that allow for an efficient energy use, e.g., in short supply systems for district heating.

The significance of *siting and zoning* can be estimated in line with urban areas as it has an essential impact on the implementation of mix of functions and density, the utilization of renewable resources, and the prevention of land use conflicts.

Furthermore, siting not marginally determines the energy demand subject to topography and exposition.

According to the function as hub for the conversion of renewable resources, the economy in rural small towns grounds on biogenic raw materials of the surrounding rural areas as *resource base*. Due to little durability and low transport densities of these resources, the transformation into commodities has to be arranged in proximity to the harvesting area ("ecology of scale"). A regional differentiation of resource potentials results in regional varieties of intermediate and manufactured products. Concerning energy generation, rural small towns can—in analogy to urban areas—mainly rely on solar energy and thermal energy recovery from waste materials of the commodity production.

Rural small towns are predestined for the implementation of district heating systems that can be operated cost- and material-efficiently if a mix of spatial functions and appropriate densities are applied. In terms of spatial energy efficiency, rural small towns enable resource and energy *"efficiency"* in the sense of the structural resilience principles. Promoting an appropriate level of density (e.g., for the dwelling function) makes a significant contribution to achieve and secure quality of life on a high level. As transport can be organized on walking and biking because of the nearness between functions, energy demand for mobility is low.

The manifestation of the resilience principle *"exposure"* is lower than in other spatial archetypes because of the relatively low energy demand and the possibilities to connect different energy domains. Furthermore, the proximity to rural areas leads to short-distance supply options which also reduce exposure.

Safeguarding and implementing mix of functions in rural small towns is in line with the structural resilience principle *"diversity"* and is aimed at enhancing the attractiveness and enlivening rural small towns. Thus, the outflow of the residential function especially to surrounding rural areas has to be prevented.

Implementing the principle of *"redundancy"* in rural small towns in particular means the interconnection of diverse energy technologies. In terms of decentralized energy supply, several small energy generation units based on regional resources can fulfill supply functions and contribute to improving the resilience against potential energy crises.

3.4 Suburban Areas

From the viewpoint of an energy-resource spatial continuum, suburban areas are dedicated to fulfill the role of an extension area and spatial reserve for urban areas as well as in a smaller amount also for suburban small towns and take over a major supply function for these three spatial archetypes, namely the provision of fresh products of daily consumption. Compared to the other spatial archetypes, suburban areas in their actual shaping can be considered furthest away from sustainability targets requiring pronounced restructuring processes. Table 3.5 illustrates a generic vision for the contributions of suburban areas to the realization of the energy turn.

Table 3.5 Generic vision of energy-optimized spatial planning for suburban areas (own table after Stoeglehner et al. 2011)

Basic function	Goal
Extension area and spatial reserve for urban areas (and suburban small towns)	Achieving highest logistic efficiency for people and goods to avoid motorized private transport
	Ensuring high living and environmental quality
	Guaranteeing basic provision of everyday goods and services, education (primary level), social and cultural services, orientation to urban areas to meet specialized requirements
Provision of food and energy for urban areas, suburban small towns, and suburban areas	Achieving highest utilization efficiency
	Reducing ecological pressures from energy provision/utilization to a minimum
	Minimizing resource consumption
Space reserve for provision of complex (industrial) goods and services	Achieving highest resource conversion efficiency
Provision of fresh goods for urban areas and suburban small towns and central hub between urban and rural areas	Achieving highest efficiency in space utilization
	Achieving maximum long-term area productivity

In suburban areas, *mix of functions* has a different meaning than in urban areas as ecological compensation, landscape-related leisure activities, and primary agricultural and forestry production are outlined more clearly. In line with the objectives of highest spatial efficiency and maximum land productivity (similar to objectives for rural areas), suburban areas should provide fresh goods for urban areas, suburban small towns, and own needs within environmental capacity limits. The residential, working and supply functions should be arranged within walking and/or biking distance. Along high-capacity public transport axes, suburban areas provide space for land-consuming industrial and commercial estates in terms of the spatial reserve for urban areas and suburban small towns (in line with the objective of maximum resource efficiency similar to urban areas).

Suburban areas provide only basic supply and are oriented to urban areas or suburban small towns concerning more specialized supply demands. Accordingly, organizing supply functions in shopping and leisure centers just outside urban areas goes in the wrong direction, especially when the dimension and location do not correspond to local needs. Such facilities consume huge amounts of arable land, demand energy, and deduct purchasing power from the centers that are subsequently confronted with decline phenomena and finally cause large amounts of motorized individual transport. To a smaller extent, this refers also to the transfer of the shopping function to settlement borders. Ecological compensation areas and recreational opportunities are to be provided for urban areas, suburban small towns, and suburban areas themselves.

Settlement *density* has to be subordinated to the basic functions as spatial reserve for urban areas as well as suburban small towns and primary production for their supply (objective of maximum spatial efficiency) and therefore be held on an appropriately high level. Concurrently, this allows for achieving the goal of highest logistic efficiency for mobility of individuals and goods.

Siting and zoning for the built environment in medium dense mixed use areas shall be oriented on high-capacity public transport axes to achieve efficiency. Complying with ideas of decentralized concentration, suburban areas as spatial reserve for complex products offer manufacturing locations for industrial and commercial sites established along regional and supraregional distribution grids (electricity, gas, heat, transport).

Resource production is considered significant, especially concerning the production of fresh goods for the provision of the urban and suburban population. Against the background of the basic function as extension area and spatial reserve, suburban areas may be facing respective land use conflicts. Autonomous energy generation is subordinate to the production of fresh products, but can build on waste materials from these processes and solar (and, if applicable wind) energy.

In view of the resilience debate, suburban areas have to face challenges especially regarding the structural principle *"efficiency"* as this spatial archetype is characterized by inefficient and spacious spatial structures. Therefore, the concentration of further urban development around high-capacity nodes of public transport systems and further increase in density result in efficiency improvements. Similar to rural areas, mobility in suburban areas can be based on alternative energy sources, e.g., electric mobility for the transport of people. Multimodal transport of goods including rail and ship leads to increase in efficiency. Due to their spacious structures, suburban areas have to face further challenges concerning the structural principle *"exposure."* Structured dismantling of undesirable spatial developments with extended land consumption can recover appropriate areas for, inter alia, primary production, ecological compensation, or recreational opportunities. Thus, the diversification of primary production and energy generation leads to an increase in *"diversity"* regarding the resilience concept. Improving environmentally friendly means of transport via a broad spectrum of routing possibilities, mobility forms and providers are one example to strengthen resilience in terms of the *"redundancy"* principle.

3.5 Suburban Small Towns

Small towns located in suburban areas can be considered as a special form of spatial archetype combining the characteristics of rural small towns in relation to the surrounding rural areas and suburban areas in relation to urban areas. A differentiation of small towns into the categories "suburban" and "rural" can be revealed inter alia from an analysis of commuter flows. Rural small towns are characterized by (1) large amounts of commuters from the surrounding rural areas and (2) commuter flows to urban areas (depending on the respective travel distances). In contrast, suburban small towns show two-way commuter flows on a high level, (1) incoming from surrounding suburban areas and also urban areas and (2) outgoing especially to the urban area toward they are oriented.

Whereas suburban areas provide basic supply, suburban small towns ensure advanced provision of goods and services that satisfy upmarket requirements. Suburban small towns can be considered as subdominant centers of the wider conurbation with amorphous settlement areas and are characterized by higher settlement densities in conjunction with housing prices on a relatively high level. Due to merging tendencies with the surrounding suburban area, external borders of suburban small towns often are not physically recognizable. The role of suburban small towns in the energy turn can be defined as illustrated in Table 3.6.

Similar to rural small towns, suburban small towns represent an attractive living space and dispose over job and supply function for the regional population. On the municipality level, *mix of functions* in suburban small towns means that basic spatial functions are located close to each other which allows for higher shares of non-motorized means of transport as cycling and walking than in monofunctional structures. On a regional level, suburban small towns are oriented toward urban areas concerning basic spatial functions that are not offered locally. Due to the location on transport axes and attractive connections to urban areas, accessibility of all basic spatial functions including inter alia education up to tertiary level is guaranteed by efficient public transport. Mix of functions leads to spatial energy efficiency as respective characteristic load functions of, e.g., dwelling and working affiliated with each other ensure a more balanced energy consumption. This results in higher efficiency of grid-bound energy systems. Furthermore, mix of functions allows for cascading use of heat as heat demands of the respective functions on various temperature levels can be satisfied.

Table 3.6 Generic visions of energy-optimized spatial planning for small towns located in suburban areas (own table after Stoeglehner et al. 2011)

Basic function	Goal
Attractive living space for decentralized industrial society/working location	Achieving and securing high quality of life
	Providing excellent leisure time opportunities
	Ensuring high environmental quality
	Guaranteeing advanced provision of goods and services that satisfy upmarket requirements, education (up to secondary level), social and cultural services
Efficient energy and resource consumption	Achieving highest utilization efficiency
	Reducing ecological pressures from energy provision/utilization to a minimum
	Minimizing resource consumption
Hub between urban and rural areas	Achieving highest efficiency in space utilization

3.5 Suburban Small Towns

Comparable to the situation in rural small towns, settlement *density* is crucial as separated functions and resulting low densities increase travel distances and make it more difficult to fulfill mobility demands by environmentally friendly means of transport. Furthermore, density is an important precondition for energy cascades as short distances between heat source and sink with the resulting minimization of heat losses are essential requirements for heat cascading.

Siting and zoning have essential impact on spatial energy efficiency as it determines the implementation of mix of functions. On a system level, the decision for a location in the small town is of greater importance than the concrete location within the spatial archetype itself.

In terms of a "resource hinterland," suburban small towns can refer to arable land in the surrounding suburban areas only to a lesser extent than small towns embedded in rural areas as the main function of primary production in suburban contexts is focussed on fresh food for the urban and suburban population. Therefore, resource conversion plays just a tangential role, whereas autonomous energy generation is of significant importance due to waste heat potentials as well as possible solar radiation harvesting areas in terms of building integrated photovoltaics related to industrial locations. Industrial waste heat and if applicable wastewater energy can be efficiently applied in cascading use (see Fig. 2.3) in efficient district heating systems as suburban small towns show adequate heat demands. Furthermore, energy cascading of waste heat from industrial and residential sites would allow for an increase in greenhouse production.

In view of the resilience concept, suburban small towns show advantages analogously to rural small towns concerning the implementation of the *"efficiency"* principle. Mix of functions and appropriate densities allow for an efficient operation of grid-bound district heating systems and contribute to the implementation of spatial energy efficiency. The arrangement of several basic spatial functions in close proximity to each other results in high quality of life and allows for the implementation of energy-efficient transport systems (e.g., walking, biking, public transport). Planning concepts as "Compact City" (OECD 2012; Jessen 1999, 2005, 2010), "City of short distances" (Brunsing and Frehn 1999), or "EcoCities" (Newman and Jennings 2008) are targeted at the implementation of spatial structures fostering such mobility forms.

As higher settlement densities are associated with higher energy intensities, suburban small towns have to face challenges concerning the resilience principle *"exposure."* Energy and resource consumption beyond own generation capacities lead to dependencies on the respective "energy and resource hinterlands." Embedded in suburban areas characterized by inefficient and spacious spatial structures and close to urban centers with high regional energy demand, suburban small towns have less area for energy generation at their disposal compared to rural small towns that can resort to the land potentials of rural areas.

Suburban small towns dispose over a broad spectrum of basic spatial functions in terms of the resilience principle *"diversity."* Oriented toward urban areas, those spatial functions that are not offered locally are accessible via well-developed public transport networks. Furthermore, the implementation of the resilience principle involves the diversification of the energy generation. Concerning the resilience principle *"redundancy,"* suburban small towns show advantages thanks to strong interlinkages with the urban areas toward which they are oriented.

3.6 Mix of Spatial Archetypes

The classic "urban–rural" dichotomy (Garreau 1991; Sieverts and Bölling 2004) is not comprehensive enough to meet the requirements of a differentiated characterization of spatial contexts (Wandl et al. 2014). In order to characterize the complex relations between the urban and the rural, territories-in-between (Wandl et al. 2014) have been introduced as new territorial classifications including *Zwischenstadt* (Sieverts 2001), *Tussenland* (Frijters et al. 2004), city fringe (Louis 1936), *Città diffusa* (Secchi 1991), territories of a new modernity (Viganò 2001), *Stadtlandschaft* (Passarge 1968), shadowland (Hamers in Andexlinger et al. 2005), spread city (Webber 1998), and *Annähernd Perfekte Peripherie* (Campi et al. 2000). Against this background, visions for energy-optimized spatial structures have to take into consideration that the previously described spatial archetypes do not only occur in their pure form but may also appear in a mix of their characteristics. This especially applies to the regional context whereby a region normally is composed of more than one of the above-described spatial archetypes.

In order to derive possible contributions of spatial structures with characteristics of several spatial archetypes to the realization of the energy turn, functional interactions between various spatial archetypes have to be taken into account. At present, commuter interdependencies contribute integral parts in delineating the hinterland of urban areas (e.g., OECD 2013; Statistics Austria, Association of Austrian Cities and Towns 2014; Wonka and Laburda 2010) that can be defined as the "worker catchment area."

Planning models such as decentralized concentration (Motzkus 2002) and polycentricism (Growe and Lamker 2012) are designed to strengthen multi-functionality, accessibility of basic and specialized supply, and promoting short distances on regional scales. In this way, functional separation and dispersed structures can be avoided, whereby increasing traffic volumes of motorized individual transport, excessive consumption of land as well as negative environmental, social, and economic impacts can be prevented. In regions and settlements of short distances, accessibility to central functions is guaranteed by means of environmentally friendly means of transport. On that account, the critical mass for utilization and efficient operation of social infrastructure and local supply can be ensured (Motzkus 2002; Krug 2005; Jessen 1999). In order to define nearness on a regional scale, Newman and Jennings (2008) suggest that all basic functions in sustainable cities have to be accessible within a 30 min walking distance and regional commuting by public transport should not take longer than 30 min.

With regard to the "energy turn" and the implementation of energy-optimized spatial structures, additional aspects have to be taken into account in terms of "energy and resource hinterlands." Several spatial archetypes claim resources that can be obtained from local or regional surrounding areas or from the global scale. At the same time, ecosystem services are required that may have impacts in local and regional but also global contexts. Considering that the availability of transport energy might be decreasing and that the "ecology of scale" might make wide range transport less feasible when the energy turn progresses, the regional scale will very likely become more important. Therefore, energy supply and resource provision as well as energy efficiency have to be regarded in a regional context (Stoeglehner et al. 2014). Furthermore, each spatial archetype offers advantages for the realization of the energy turn, may it be with extended energy efficiency potentials or a wider range of resource and energy provision options, and can contribute to the implementation of the spatial energy resilience principles in different ways. Certain disadvantages depending on the spatial characteristics are associated with each archetype that may be compensated by the respective strength of other spatial archetypes in a regional context. Intensive interchange and integration of the respective potentials in terms of spatial and energy aspects on the regional level and the compliance with the resilience principles allow for an effective implementation of the spatial energy resilience concept.

References

Aguilar, A. G., & Ward, P. M. (2003). Globalization, regional development, and mega-city expansion in Latin America: Analyzing Mexico City's periurban hinterland. *Cities, 20*(1), 3–21.

Andexlinger, W., Kronberger, P., Mayr, S., Nabielek, K., Ramière, C., & Staubmann, C. (2005). *TirolCity*. Vienna: Folio Verlag.

Association of Austrian Cities and Towns. (2015). *Smart Cities: Menschen Machen Städte*. Tätigkeitsbericht des österreichischen Städtebundes. Wien: Verlag für moderne Kommunikation.

Benediktsson, J. A., Pesaresi, M., & Amason, K. (2003). Classification and feature extraction for remote sensing images from urban areas based on morphological transformations. *IEEE Transactions on Geoscience and Remote Sensing, 41*(9), 1940–1949.

BEV – Bundesamt für Eich- und Vermessungswesen (2015). Stichtagsdaten der Verwaltungsgrenzen zum 01 Apr 2015.

Brown, L. A., & Holmes, J. (1971). The Delimitation of Functional Regions, Nodal Regions, and Hierarchies by Functional Distance Approaches. *Journal of Regional Science, 11*(1), 57–72.

Brunsing, J., & Frehn, M. (1999). *Stadt der kurzen Wege. Zukunftsfähiges Leitbild oder planerische Utopie?*. Dortmund: Informationskreis für Raumplanung.

Campi, M., Bucher, F., & Zardini, M. (2001). *Annähernd perfekte Peripherie. Glattalstadt/Greater Zurich Area*. Basel: Birkhäuser.

Creutzig, F., Baiocchi, G., Bierkandt, R., Pichler, P. P., & Seto, K. C. (2015). Global typology of urban energy use and potentials for an urbanization mitigation wedge. *Proceedings of the National Academy of Sciences of the United States of America, 112*(20), 6283–6288.

Douglass, M. (2000). Mega-urban Regions and World City Formation: Globalisation, the Economic Crisis and Urban Policy Issues in Pacific Asia. *Urban Studies, 37*(12), 2315–2335.

Exner, A., Politti, E., Schriefl, E., Erker, S., Stangl, R., Baud, S., et al. (2016). Measuring regional resilience towards fossil fuel supply constraints. Adaptability and vulnerability in socio-ecological Transformations – the case of Austria. *Energy Policy, 91*, 128–137.

Frijters, E., Hamers, D., Johann, R., Kürschner, J., Lörzing, H., Nabielek, K., et al. (2004). *Tussenland.* Rotterdam: NAi Uitgevers.

Garreau, J. (1991). *Edge city: Life on the new frontier.* New York: Doubleday.

Growe, A., & Lamker, C. (2012). Polyzentrale Stadtregion – die Region als planerischer Handlungsraum. In: A. Growe, K. Heider, C. Lamker, S. Paßlick, Th. Terfrüchte (Eds.), *Polyzentrale Stadtregionen – Die Region als planerischer Handlungsraum.* Arbeitsberichte der ARL 3. Hannover.

Hartshorne, R. (1933). Geographic and Political Boundaries in Upper Silesia. *Annals of the Association of American Geographers, 23*(4), 195–228.

Hidle, K., Farsund, A. A., & Lysgard, H. K. (2009). Urban-Rural Flows and the Meaning of Borders. Functional and Symbolic Integration in Norwegian City-Regions. *European Urban and Regional Studies, 16*(4), 409–421.

Jessen, J. (1999). Stadtmodelle im europäischen Städtebau – Kompakte Stadt und Netz-Stadt. In: Becker, H., Jessen, J., Sander, R. (Eds.): *Ohne Leitbild? Städtebau in Deutschland und Europa.* (pp.490–504). Stuttgart: Karl Krämer Verlag.

Jessen, J. (2005). Leitbilder der Stadtentwicklung. In: Akademie für Raumforschung und Landesplanung (Ed). *Handwörterbuch der Raumordnung.* 4., neu bearbeitete Auflage (pp. 602–608), Hannover: Verlag der ARL.

Jessen, J. (2010). Leitbilder der Stadtentwicklung und des Städtebaus. In: Bott, H., Jessen, J., Pesch, F. (Eds.). *Lehrbausteine Städtebau – Basiswissen für Entwurf und Planung.* 6., grundlegend überarbeitete Auflage (pp. 121–128). Stuttgart: Universität Stuttgart Fakultät für Architektur und Stadtplanung.

Krug, H. (2005). *Räumliche Wahlmöglichkeiten als Effizienzkriterium für Siedlung und Verkehr.* Szenarien – Modellrechnung – Vergleichende Bewertung. Dissertation an der Universität Kassel.

Louis, H. (1936). *Die geographische Gliederung von Gross-Berlin.* Stuttgart: Engelhorn.

Motzkus, A. -H. (2002). *Dezentrale Konzentration - Leitbild für eine Region der kurzen Wege?* Auf der Suche nach einer verkehrssparsamen Siedlungsstruktur als Beitrag für eine nachhaltige Gestaltung des Mobilitätsgeschehens in der Metropolregion Rhein-Main. Bonner Geographische Abhandlungen 107. Sankt Augustin: Asgard.

Newman, P., & Jennings, I. (2008). *Cities as Sustainable Ecosystems: Principles and Practices.* Washington, DC: Island Press.

OECD. (2012). *Compact City Policies: A Comparative Assessment, OECD Green Growth Studies.* Paris: OECD Publishing.

OECD. (2013). Definition of Functional Urban Areas (FUA) for the OECD metropolitan database. http://www.oecd.org/gov/regional-policy/Definition-of-Functional-Urban-Areas-for-the-OECD-metropolitan-database.pdf. Accessed 09 Nov 2015.

Passarge, S. (1968). *Stadtlandschaften der Erde.* Hamburg: Cram, de Gruyter.

Pötscher, F., Winter, R., & Lichtblau, G. (2010). *Elektromobilität in Österreich. Szenario 2020 und 2050.* Report REP-0257, Vienna.

Rashed, T., Weeks, J. R., Roberts, D., Rogan, J., & Powell, R. (2003). Measuring the physical composition of urban morphology using multiple endmember spectral mixture models. *Photogrammetric Engineering and Remote Sensing, 69*(9), 1011–1020.

Secchi, B. (1991). *La periferia.* Casabella 583.

Seto, K. C., Dhakal, S., Bigio, A., Blanco, H., Delgado, G. C., Dewar, D., et al. (2014). Human settlements, infrastructure and spatial planning. In: O. Edenhofer, R. Pichs-Madruga, Y. Sokona, E. Farahani, S. Kadner, K. Seyboth, A. Adler, I. Baum, S. Brunner, P. Eickemeier, B. Kriemann, J. Savolainen, S. Schlömer, C. von Stechow, T. Zwickel, J. C. Minx (Eds). *Climate Change 2014: Mitigation of Climate Change. Contribution of Working Group III to the Fifth Assessment Report of the Intergovernmental Panel on Climate Change.* Cambridge and New York: Cambridge University Press.

References

Sieverts, T. (2001). *Zwischenstadt zwischen Ort und Welt, Raum und Zeit, Stadt und Land.* Bertelsmann Fachzeitschriften, Gütersloh, Berlin: Bauverlag.

Sieverts, T., & Bölling, B. (Eds.). (2004). *Mitten am Rand. Auf dem Weg von der Vorstadt ueber die Zwischenstadt zur regionalen Stadtlandschaft.* Wuppertal: Verlag Müller + Busmann KG.

Statistics Austria. (2016). Population stock on 1st January 2015 on municipal and settlement level. Vienna.

Statistics Austria. (2012). Urban regions delineation 2001—ordered by municipalities. http://www.statistik.at/wcm/idc/idcplg?IdcService=GET_PDF_FILE&RevisionSelectionMethod=LatestReleased&dDocName=065037. Accessed 09 Nov 2015.

Statistics Austria, Association of Austrian Cities and Towns. (2014). Österreichs Städte in Zahlen 2014. http://www.statistik.at/wcm/idc/idcplg?IdcService=GET_NATIVE_FILE&RevisionSelectionMethod=LatestReleased&dDocName=080540. Accessed 09 Nov 2015.

Stoeglehner, G., Erker, S., & Neugebauer, G. (2014). *Energieraumplanung.* Materialienband. In Zusammenarbeit mit der ÖREK-Partnerschaft "Energieraumplanung". ÖROK Schriftenreihe Nr. 192. Wien: Bundesministerium für Land- und Forstwirtschaft, Umwelt und Wasserwirtschaft, Geschäftsstelle der Österreichischen Raumordnungskonferenz (ÖROK).

Stoeglehner, G., Niemetz, N., & Kettl, K.-H. (2011). Spatial dimensions of sustainable energy systems: new visions for integrated spatial and energy planning. *Energy, Sustainability and Society, 1*(2), 1–9.

Stremke, S., van den Dobbelsteen, A., & Koh, J. (2011). Exergy landscapes: Exploration of second-law thinking towards sustainable landscape design. *International Journal of Exergy, 8*(2), 148–178.

Un, D. E. S. A. (2015). *World Urbanization Prospects: The 2014 Revision.* New York: United Nations Department of Economics and Social Affairs, Population Division.

Viganò, P. (2001). *Territori della nuova modernità. Il piano territorial di Lecce.* Napoli: Elect Napoli.

Wandl, A., Nadin, V., Zonneveld, W., & Rooij, R. (2014). Beyond urban-rural classifications: Characterising and mapping territories-in-between across Europe. *Landscape and Urban Planning, 130,* 50–63.

Webber, M. M. (1998). The joys of spread-city. *Urban Design International, 3*(4), 201–206.

Wonka, E., & Laburda, E. (2010). Stadtregionen 2001 – Das Konzept. http://www.statistik.at/wcm/idc/idcplg?IdcService=GET_PDF_FILE&RevisionSelectionMethod=LatestReleased&dDocName=058285. Accessed 09 Nov 2015.

Chapter 4
Fields of Action for Integrated Spatial and Energy Planning

Gernot Stoeglehner, Michael Narodoslawsky, Susanna Erker, and Georg Neugebauer

Abstract This chapter shows how integrated spatial and energy planning is connected to further issues relevant to climate protection and the energy turn. It is discussed how consistent decision making can be organized and which actors have to be addressed in implementing integrated spatial and energy planning. As core of this chapter, 7 normative principles for integrated spatial and energy planning are elaborated in the three core fields of action: (1) energy-efficient spatial structures, (2) renewable resources and spatial structures, and (3) energy supply systems tailored to spatial structures.

Designing the energy turn poses major challenges for shaping planning processes that integrate spatial structures and energy systems. On the one hand, these challenges arise from the high complexity of the systems to be integrated, as already described in Sect. 4.2 of this book. On the other hand, the social process of designing and implementing the energy turn is highly complex, bringing together multiple actors and stakeholders with a high diversity of interests, sometimes pointing in totally different directions.

Adding to this complexity, energy turn and climate protection are embedded in a network of framework conditions and influencing factors at the local, regional, national, and supranational scale, as pointed out in Fig. 4.1. The issues described in Fig. 4.1 as framework conditions and influencing factors—besides integrated spatial and energy planning—can be detected in the following domains (Stoeglehner et al. 2014):

- Value base of society: It reflects the question, how important energy turn and climate protection are for society, and how much support and democratic legitimation and support is to be expected for integrated spatial and energy planning in favor of the energy turn and climate protection.
- Policies and administrative frameworks: define the legislative and executive power of societies which design and implement laws in all fields of governmental action including energy, economy, taxation, environment, spatial planning, and regional development.

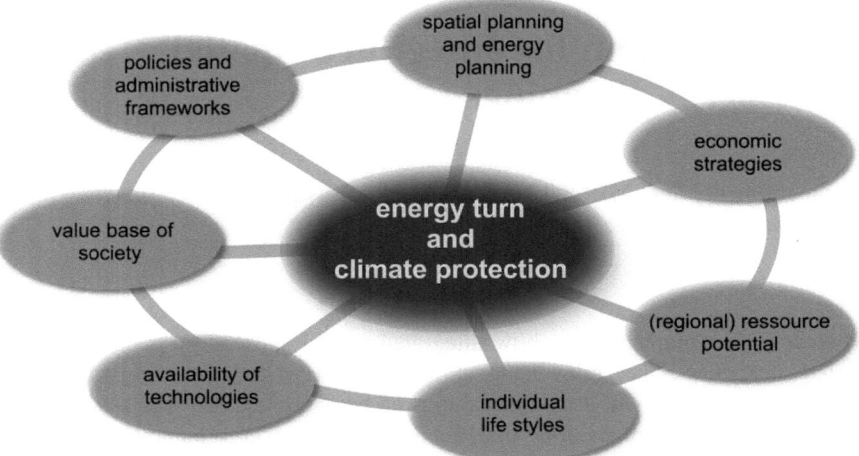

Fig. 4.1 Influencing factors for integrated spatial and energy planning (own illustration after Stoeglehner et al. 2014)

- Economic strategies: On the macro-level, they include governmental actions such as public investments, framework design for private sector investments, and public–private partnerships; on the mircolevel, they comprise behavioral patterns and economic practices of public bodies acting as private sector players, companies, and households.
- (Regional) resource potentials: Both private and public sector decisions are bound to the resource potential that can be tapped for the implementation of policies or individual (economic) strategies. The resource potential includes natural resources, e.g., renewable energy sources, human resources such as the knowledge embedded in local and regional societies, or the social capital of a region, i.e., the quality of relations within a society.
- Individual lifestyles: They reflect the individual values, which are linked to the value base of society, and their application in everyday decisions about the organization of everyday life, consumer behavior, mobility choices, preferred spatial structures for dwelling, recreation, etc. They are also influenced by the economic power of households or availability of technologies and heavily influence the resource and energy intensity of private life.
- Availability of technologies: Availability of technologies can be a matter of research and development, but also a matter of prices or regulatory frameworks. For instance, individual car transport massively shapes urban and rural areas. Even most of the spatial developments of the past 70 years would not have been feasible or even imaginable without the broad availability of cheap, almost not or little restricted individual transport both concerning persons and goods. Anyhow, availability of technologies in transport sectors, building and

construction sectors, the energy sector, for consumer products, information, and communication technologies, etc., are an enormous influencing factor for the design and implementation of the energy turn and climate protection.

Integrated spatial and energy planning is one of the nodes in this network, which is closely linked to all other nodes and reciprocally influenced by other issues as the other issues are influenced by integrated spatial and energy planning. For example, the structures created by integrated spatial and energy planning shape individual behavior and economic practices, as they also affect the spatial structures, all of these issues reflecting the value base of society.

For instance, in Austria, the desired dwelling form for the majority of people is a single-family house. If the economic power of households is strong enough, the fulfillment of this "dream" shapes spatial structures as more single-family houses are built. The same is true with economic strategies and practices of companies: "Just in time" production and delivery poses not only challenges to logistics, but has enormous influences on the design of industrial areas, the need for transport infrastructure and the transport-related environmental impacts.

Within these influencing factors according to Fig. 4.1 as boundary conditions, a multitude of actors have to be coordinated and motivated to act in support of the energy turn. As can be derived from the broadness of the influencing factors, they range from the supranational government level, e.g., when it comes to climate protection targets, to the individual level of households and companies. Looking at the diversity of actors in more detail, six groups of actors including "framework designers," "lobby organizations," "know-how developers," "planning community," "developers, investors, and operators" as well as "end users" can be identified. Figure 4.2 shows this network of actors relevant for the energy turn and climate protection as ascertained in an intergovernmental working group dealing with issues of integrated spatial and energy planning in Austria (Stoeglehner et al. 2014).

This high diversity of stakeholder groups, the value base of society, but also the value base of interest groups or individuals, poses great challenges to draw consistent frameworks in policy design, not only related to the energy turn or climate protection, as can be described with the "decision-making pyramid" according to Stoeglehner and Narodoslawsky (2008). From a sustainability perspective, a decision related to one domain of sustainability very likely has considerable effects on the others, e.g., an environmental decision on social and economic issues, and vice versa (see Fig. 4.3): When it comes to implementation, the shift of decision-making levels from the (supra-)state to the individual level normally means a shift in perspectives looking at a certain problem.

Climate protection policies are clearly dominated by an environmental perspective, while already the energy turn widens the still prevailing environmental approach with issues of economic feasibility, supply security, or resilience in light of energy, global political, or economic crises. Yet, the implementation on the grassroot level, the individual company or individual household, is mostly not an

framework designers
- decision makers designing policies and funding systems on international, European, national and regional level
- legislative bodies at state and federal level, both in core fields of spatial and energy planning as well as related laws like, inter alia, tenancy law, property rights, building regulations, construction law or parking space regulations
- law enforcement authorities on national, regional and local levels, e.g. spatial planning authorities, energy planning authorities, environmental protection authorities,

lobby organisations
- organised groups of agents that influence policies, legislation, enforcement and planning
- NGOs, project opponents, media, educational institutions, neighbours, residents and landowners

know-how developers
- research and development concerning renewable energy, energy efficiency, energy storage and distribution (e.g. smart grids), energy system design
- research and development concerning different planning domains like spatial and environmental planning, infrastructure planning including mobility etc.
- research and development in further related sectors like building and construction sectors, information and communication technologies etc.

planning community
- governments responsible for spatial planning on all relevant levels, e.g. national or regional levels
- municipalities in the role of planning authorities, parties or planning agencies
- community organisations, local or regional non-governmental organisations with an interest in planning
- freelance planners

developers, investors and operators
- actors with economic interests in the implementation of projects
- mobility and infrastructure providers and sustainers, site developers and building land acquisition funds
- planning agencies, project developers, energy suppliers, network operators, industry as a producer of energy technologies, investors

end-users
- population and economy, representing lifestyles and business practices
- people who influence the implementation of the energy turn by everyday decisions of households and companies

Fig. 4.2 Actors and stakeholders relevant for integrated spatial and energy planning (own illustration after Stoeglehner et al. 2014)

environmental decision, but an economic or societal decision. For instance, if a greenhouse gas reduction policy concerning the emissions of room heating is successful, finally depends on the choices of companies or households when they decide for a new heating device and/or thermal insulation. While a pure economic decision framework might ask for the cheapest possibility, normally a trade-off between these issues takes place, e.g., considering an affordable means of heating in

Fig. 4.3 The "decision-making pyramid" (own illustration after Stoeglehner and Narodoslawsky 2008)

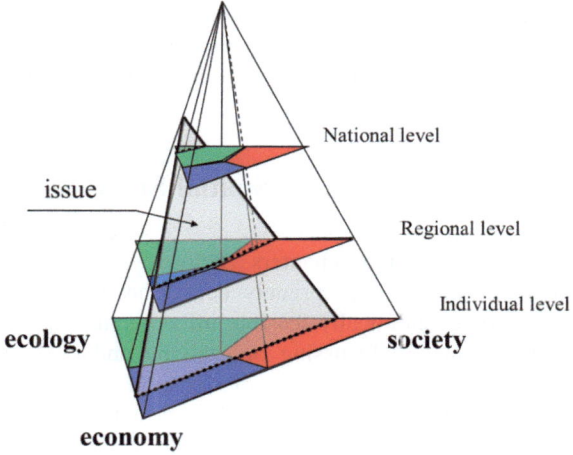

light of "environmental friendliness" or social prestige or prejudice, e.g., about the perceived image a company or household might generate when choosing a means of room heating. Therefore, consistent policy design concerning the energy turn and climate protection means to draw frameworks which allow grassroot-level actors to make an environmentally friendly decision, even though economic or social issues might be in the center of decision-making.

In principle, governments have four possibilities to interact with societies: (1) the creation and enforcement of legal norms and taxation; (2) subsidies in order to set economic incentives for individual behavior in favor of certain policies; (3) public sector investments or investments in public–private partnerships in line with certain policies; and (4) the raising of awareness for certain policies. Depending on the group of actors and the kind of decisions, different action channels are necessary. In this chapter, we mainly address governmental action in designing consistent frameworks to implement the energy turn and climate change protection with means of integrated spatial and energy planning. In Chap. 5, we will discuss these issues with respect to individual planning processes on the local and regional level and their respective key actors. Therefore, for this section, actors such as the framework designers themselves, lobby organizations, and know-how developers are of special interest, and for Chaps. 5 and 6, it will be the planning communities, developers, investors, and operators as well as end users. The latter are important target groups for the design of frameworks as well, but have limited capacity to perform as actors in framework design processes—as long as they have no lobbying power. In the following subchapters, three fields of action are introduced, drawing a short vision based on the explanations of Chaps. 2 and 3, and discussing the main challenges and the cornerstones for policy implementation as principles of integrated spatial and energy planning with respect to diverse spatial structures. Therefore, this chapter is normative, with the value frame linked to sustainable spatial development. Spatial structures are understood as physical agglomerates of several elements of the same kind or of different kinds which are connected by functional relations, may these elements be, for instance, buildings, open spaces, infrastructures, or parts of open (cultural) landscapes.

4.1 Energy-Efficient Spatial Structures

Summing up the chapters before, energy-efficient spatial structures are multi-functional, appropriately dense, have a certain minimum size in order to guarantee economically feasible infrastructure and energy supplies, and are compactly organized focussing on walking and cycling in combination with public transport as primary means of transport. In Chap. 3, we laid out how energy efficiency of spatial structures should be achieved in different spatial archetypes and their combination. Therefore, Principle 1 can be directly derived from the chapters before.

Principle 1: Create multi-functional, appropriately dense and compact spatial structures

Although this claim is in line with many planning visions as already elaborated in the previous sections, and, therefore, not new to spatial planners, recent spatial developments point in the opposite direction. Creating such spatial structures would allow for multiple win-win-win situations, e.g., an efficient use of building land, which should make the aligned land uses as well as the infrastructure provision more economically efficient (Dallhammer and Mollay 2008), an easier organization of everyday life, so that for instance the labor force participation rate of women is higher in compact spatial structures than in sprawl situations (Baaske 2013), less environmental pollution (Newman and Jennings 2008), and many more. It seems that the energy turn would just provide another line of argument to promote such settlement structures.

Yet, providing energy-efficient settlement structures poses a major, at the moment seemingly, irresolvable challenge to spatial planning. Actual developments show sprawl, land consumption for infrastructures as well as settlements, commercial, and industrial areas, as global phenomena. How does that come, even though many policy papers and strategy papers for spatial planning point in the right direction, which means that the value base of society might be even right? The framework shown in Fig. 4.1 provides some starting points for explanations.

For instance, policies and administrative frameworks are often organized in sectors. Solutions, which make sense in one sector, are less attractive when intersectorial aspects are taken into consideration. Non-holistic, sector-oriented policies, and administrative practices might lead to wrong steering from an overall perspective. Examples are manifold, e.g., subsidies or taxation that are equally distributed without taking the spatial structure into account, economic policies favoring the same types of projects in all types of spatial structures, even spatial planning regulations that promote a strict separation of housing and economic activities due to emission protection reasons, the siting of public facilities sometimes also does not reflect the design principles of mix of functions, density, compactness and accessibility with public transport etc.

It can be observed that economic strategies of companies and public bodies are often related to short-term thinking. For choosing locations, companies normally look for certain spatial qualities concerning the infrastructure supply, availability of areas, possibilities to develop, etc. These considerations often call for greenfield developments, which are easier, expecting less conflict with neighboring land uses, and often also being cheaper than sites in central locations. Also, economic decisions of landowners and developers favor the financially most profitable land uses in the central areas of cities and towns, pushing the land prices so that low-to-medium income dwellers including the middle-class and small businesses tend to be pushed out of the qualitatively high core zones of localities. In the long run, such developments very likely lead to a decrease of the intensity and/or diversity of land uses and a decline of the spatial quality. Cities and towns are vivid and lively if a high diversity of dwellers, employees, workers, customers, etc., uses them throughout the times of the day. Focusing too much on certain land uses and their promise of profit for the landowners and developers excludes a number of user groups and leads to undesired urban development in the long run, e.g., by pushing inhabitants, certain economic sectors, and jobs as well as daily supply out of the urban centers. A high diversity of land uses and respective user groups is in the long-term interest not only of the public domain, but also of the users of the spatial structures as well as the landowners and developers. This kind of holistic thinking is not developed everywhere and would, for instance, mean to maintain, re-establish, or introduce social housing in urban centers in order to guarantee for dwellings in all price categories and a high diversity of inhabitants.

Individual lifestyles are volatile over a lifetime. Spatial structures are persistent. Many people prefer, for whatever reasons, a single-family house as dwelling. This form of living has some advantages, especially for young families with children, but the qualities of single-family houses can also be achieved in more dense settlement structures. Yet, it is easy to find low-quality, high dense settlements that spur negative prejudice against denser ways of urban and settlement design. Furthermore, single-family housing also has considerable disadvantages, such as a lack of efficient public transport or daily supply, longer distances to cover, dependence on cars, and the duty to look after house and garden, which might pose negative impacts on quality of life and individual resilience in different phases of life. As much individual capital (often borrowed) is bound to the decision for a dwelling, taking such considerations into account is very important when choosing the kind of and location for the dwelling as center of life. Therefore, a lot of awareness rising is necessary to make people fully understand the consequences their choice of a dwelling has, and to allow them to make decisions in their best interest.

Only this short reflection shows that policies and administrative practices in the governmental sphere have to be evaluated and revised in order to implement Principle 1 and reach relevant target groups for this implementation. As a solution to this challenge, a framework of consistent governmental action has to be defined that rests on the coordination of the four possibilities of governmental action, which are legislative interventions, financial incentives, public investments, and awareness rising. In order to succeed in this coordination, spatial analysis can provide the fundamental information about and appraisal of desired and undesired consequences for spatial development in the spatial archetypes described in Chap. 3. This kind of impact appraisal can be used as a basis for the drafting of respective policies. Concerning the planning tasks at hand, spatial planning frameworks can contribute clear criteria for the zoning of the desired development areas. This leads us to Principle 2.

Principle 2: Define core areas for the future development of energy-efficient spatial structures

As the creation of energy-efficient spatial structures has to be related to the characteristics, limitations, and potentials of certain spatial archetypes, also the coordination of different policy fields has to be bound to spatial characteristics. First of all, core areas for the future development of energy-efficient spatial structures have to be defined. In the core areas of settlement structures, Principle 1 should be primarily implemented. In core areas, multi-functional, appropriately dense, and compact built structures with sufficient open space supply should be provided, e.g., by redevelopment of brownfields, densification of existing built structures, and supplementing of missing spatial functions.

We define these core areas taking two levels of detail into account (after Stoeglehner et al. 2013): (1) the question, which localities as connected areas of building land (city, town, or village cores) can be suitable; and (2) the question, which areas within the localities would be applicable as core areas.

To define, which localities would be suitable to accommodate a core area of future spatial development, we suggest the following criteria (after Stoeglehner et al. 2013): (1) a certain size of the locality, (2) a minimum density, (3) a minimum standard of public infrastructures and supply with goods and services, and (4) if applicable, a historic settlement core. Size is important to guarantee a certain demand for public infrastructures and buying capacity, so that daily as well as more specialized goods and services can be offered. Density is a measure of efficiency concerning both energy and public infrastructures and keeps distances shorter. The minimum requirements for infrastructure and supply provide for a minimum quality of life and reduce transport demands. Finally, in many cities, towns, and villages, the revitalization of historic settlement structures is a goal of spatial planning and regional development, so that this criterion can be taken into consideration independently from the actual use of such historic structures. We propose that if such criteria are met—no matter how the area is used at the moment—redevelopment of these areas according to Principle 1 is possible, improving the quality of life of the present users, building on existing structures, and setting incentives for future sustainable spatial development. From an energy supply perspective, such developments help to increase energy efficiency and, inter alia, reduce individual motorized transport.

Within the localities determined according to the above-mentioned criteria, core areas of development can be defined by the following criteria: (1) land use zoning that allows a mix of residential, public, and private services, educational, and health-care facilities as well as non-emitting manufacturing trade; (2) several story buildings to secure density and mix of functions within the buildings; (3) provide at least 30–50 % of the floor space for dwellings—only settlement cores are vivid around the clock if people permanently live there; (4) a bigger floor height at least in the ground floor to allow for different kinds of functions; and (5) supply facilities within walking distance.

Depending on the spatial archetypes, the localities are situated, as well as considering the value base of society, the criteria should be quantified, e.g., the amount of inhabitants for a sufficient size, the definition of density, or the questions, what kind of facilities should be included in daily supply or how far walking distance is. Furthermore, the definition of a historic settlement core has also to be negotiated, determining criteria such as, the age of the buildings or the layout of streets and buildings referencing to certain settlement epochs independent of the actual building age.

In particular, in rural areas, a downsizing of criteria is necessary, as according to Chap. 3, basic supply has to be granted although densities are normally low and mix of function is reduced to only a few. In order to take these circumstances into consideration, core areas could be determined by radii around supply facilities (e.g., kindergarten, school, shop, and public transport) and maximum distances between them, oriented on walking for less mobile population groups such as children and elderly people, and the availability of public transport. Also, the minimum quality for public transport has to be defined.

Practice in many countries shows that such spatial planning provisions do not countervail against undesired spatial development such as sprawl, inter alia, because of the reasons discussed in Principle 1. Furthermore, such regulatory frameworks work better in prohibiting undesired developments than in enforcing favored developments. In order to discharge such regulations with power, further measures complementing spatial planning ideas have to be taken (after Stoeglehner et al. 2011, 2013, 2014):

- Legal interventions: In order to achieve a consistent regulatory framework, certain laws have to be revised and adjusted. Such legal amendments have different points of departure, such as:
  - Property laws: Core areas have to be made available for the development. Therefore, it has to be discussed with which measures these—from a public interest perspective—highly valuable development areas should be made available for the desired spatial structures and the desired social diversity. The range of measures could reach from certain limits for rental prices to the obligation for landowners to implement the planned developments (with or without compulsory purchase). Such obligations can also be combined with financial incentives. If space in core areas is not available, developments will be replaced to the second or third best locations, which is against public interest and the interests of the affected users.

- Emission laws: Certain emission protection laws—which have been drafted in spirit of separation of functions—should be revised. There are some land uses in the production sector that should be separated from others, but especially, when it comes to the mix of housing, service, and retail sectors as well as public facilities and some non-disturbing production branches (e.g., zero-emission-manufacturing), more flexibility to promote a mix of functions would be favorable to create energy-efficient spatial structures, especially in core areas of development.
- Taxation: Different levels of taxation would be appropriate to enforce energy-efficient spatial structures; for example, the property tax could be bound to mix of functions and density, asking for the lowest tax level for the desired mixed-function areas with medium densities. If densities get to high (mainly in urban centers) or too low (mainly in rural or suburban areas), the property tax could increase. Furthermore, on unused or underused parcels of land in core areas, higher taxes should be imposed.
- Construction law: In construction law, certain provisions could be implemented concerning the promotion of a mix of function, e.g., bigger height between floors to allow for commercial, office, or public uses. Also, the amount of car parking spaces could be regulated related to the respective settlement structure.
- Transport sector: Different restrictions for the individual motorized transport sector could be imposed, both concerning the transport of persons and goods. This could start with parking regulations, car-free zones, etc., and should also allow for creating attractive streets for non-motorized individual transport.

- Financial incentives: Drawing the line around core areas of energy-efficient spatial structures is especially effective if certain subsidies are only granted within these borders. These subsidies should include the following:

  - Housing subsidies: In particular, apartment buildings should be focused in core areas of spatial development, allowing for a mix of functions even within the building, e.g., by retail, office, or public facilities in the lower floors and dwelling in the upper floors.
  - Economic subsidies: Also, businesses subsidies, if granted, should be attributed to spatial qualities, meaning that in the retail and service sector, developments should be only allowed in core areas. Production sectors and logistics have different site requirements and emission patterns, so mix of functions has to be differently interpreted: First, they should be located on nodes of different grids, e.g., transport grids, connecting different means of transport including railways, and energy grids such as gas, electricity, district heat, and cooling. Second, they should be in due nearness to settlements to allow for the use of waste heat in district heating and cooling systems of core areas of development, and to organize personal transport between core areas and industrial zones by walking, cycling, or public transport.

4.1 Energy-Efficient Spatial Structures

- Real estate: Certain subsidies should be granted or public bodies or funds established as players on real estate markets in order to mobilize building land in core areas, e.g., by buying underused property.
- Commuter subsidies: In countries such as Austria and Germany, commuting is subsidized. Such subsidies should only be granted for public transport and for housing sites located in rural core areas of settlements.

- Public sector investments: Public sector investments including public–private partnerships give governments the chance to create role models of good and best practice. They also show how serious governments take their own energy turn and climate protection policies in building public infrastructures. This includes the choice of sites for social infrastructures, e.g., hospitals, schools, and administrative buildings, which should be located in core areas of energy-efficient settlement structures, energy-efficient building standards, or the amount and relation of infrastructure investments for walking, biking, public transport, and car transport.
- Awareness rising: This is a core task, as it shall create the understanding and acceptance for certain measures by the affected population and economic actors. This understanding and acceptance is very important as it effects the value base of society and the enforceability of action plans implementing any kind of policy of governments that are democratically legitimized—and want to be re-elected. Furthermore, awareness rising influences on lifestyles and economic practices at an individual scale. Inter alia, the question about the affordability of renewable energy supplies can be impaired, changing to "what is cheapest?" to "what can I afford taking environmental protection and social responsibility into account?" or even to "do I need this investment?" As long as economic choices in favor of the energy turn, climate protection, and integrated spatial and energy planning are not the cheapest ones, this shift in perspectives, considering sufficiency, is crucial and needs a lot of awareness building and convincing to lead to corresponding action.

Finally, any combination of these measures is necessary as it aggravates the impacts of each strategy. Awareness is always important, not only because of the change of everyday action, but also because it creates societal legitimation, understanding and support for legal measures, and the change of financial incentives and public investments.

4.2 Renewable Resources and Spatial Structures

In Chaps. 2 and 3, we have already discussed that the use of renewable energy resources is critically dependent on the spatial context. This leaves spatial planning with the delicate task to include complex issues of resource logistics, optimal siting of energy conversion technologies, and different spatial aspects of energy distribution grids into the already complex planning process. This task becomes even more daunting as not only spatial aspects but also temporal availability and natural aspects have to be factored in. The following principles can be seen as guidelines for this delicate balancing act.

Principle 3: Define shares of renewable energy resource use based on local and regional environmental capacity limits, considering social responsibility and concurring economic uses

Any planning for the energy turn on the local and regional level must start with taking stock of the available renewable resource basis. The challenge here is, however, that the available resource basis is not a fixed entity that may easily be calculated, analyzing the map and environmental parameters of a given area. Figure 4.3 clearly shows the interconnectedness of environmental, social, and economic decisions that form the basis of the transition to a low carbon, sustainable energy system. The previous chapters also reiterated the fact that the energy turn will spur competition for land that has to be resolved in an open and fair societal planning process. Defining the share of renewable resources used for energy as well as other purposes is the first step in this planning process, which requires a comprehensive public consultation process along the principles laid down in Chap. 28 of the UN AGENDA 21 (United Nations 1992) and the follow-up documents (United Nations 2016).

Although this consultation process must be tailored to the political, societal, and cultural framework in the region, there are some topical issues that have to be addressed in this process. These are as follows:

- Land has many functions, from providing resources to preserving vital natural cycles such as the water cycle to ensure diversity of species. Without consensus on the relation between these functions in the concrete region, no resource planning is possible.
- All regions are interconnected, and there exists no such thing as "regional autarky." This requires reflecting about the responsibilities to supply other regions with resources generated on the land (e.g., food or energy) but also on the level of dependency from other regions. Rather than striving for autarky, this means to identify those functions of land use that the region wants to specifically develop as a basis for its own wealth and its role in the concert of regions.
- Preservation or improvement of functionality of land as a purveyor of natural services and resources must be the solid basis for sustainable land use. This becomes more important when competition for these services increases as food sector, process industry, and energy sector, not to mention tourism and nature preservation vie for the same land. Preserving natural production factors such as soil quality and water availability is however the result of social, economic, and technological decisions that have profound impact on resource availability. This includes decisions about agricultural practices, the fraction of harvest residues retained for soil management, and the ability to close material cycles.

Landscapes play a particular role for the identification of citizens with their region as well as for economic activities such as tourism and the leisure industry.

Utilization of renewable resources may have considerable impacts on landscapes (as well as cityscapes). The consultation process therefore must also address the issue of how much and what change in landscape is tolerable in the region.

Principle 4: Zone priority and exclusion areas for the use of renewable resources taking spatial structures and land use conflicts into account

The use of renewable resources is not necessarily sustainable as already pointed out above, requiring societal decisions about the level of desired positive or tolerable negative environmental, social, and economic impacts of the broad application of renewable resources and related technological choices. Spatial planning cannot only provide processes to negotiate the aspired or acceptable level of impacts; it may also provide planning instruments to secure areas, where renewable resource provision is prioritized to other land uses, or to exclude certain renewable resource uses in specific areas in order to prevent land use conflicts.

The first measure toward securing large areas for renewable resource use is to implement the concept of energy-efficient spatial structures. Multi-functional, dense, and compact settlement structures save not only energy, but also soil, and prevent sprawl. Therefore, more contiguous, large cultural landscapes can be secured where renewable resources can be utilized for energy provision. For instance, many land use conflicts around wind power plants can be resolved with safety distances, where safety distances and size of wind parks are proportional (Felber and Stoeglehner 2014). The larger cultural landscape areas between settlements are, the more likely large-scale wind power generation will find suitable sites. Areas especially suitable for certain energy generation technologies—no matter which one—should be left free from conflicting land uses, which can be done by priority areas on the supraregional, regional, and local scale.

Furthermore, certain energy generation purposes should be excluded from certain areas, e.g., hydropower from sensitive riverine landscapes with high biodiversity and nature protection value, or wind power parks from bird migration routes. Transformation of landscape sceneries might be a sensitive topic in tourism regions. Therefore, exclusion zones can mitigate land use conflicts by forbidding renewable resource utilization, giving other land uses priority in such areas. The question, how many of the priority and exclusion areas are needed, also depends on the agreement about the value base related to the energy turn, as the targeted energy demand, the aspired energy technology mix for generation, distribution, and storage as well as the potential to export energy from the region or locality under survey. The acceptable consequences of certain energy technologies, as well as the no-go-areas for energy generation, distribution, and storage have to be determined. Therefore, to plan priority and exclusion areas is a highly complex task.

Finally, this task has to be concretized to the level of siting for renewable energy supply facilities within the priority and exclusion zones, taking potential impacts on the micro-scale into account. If priority and exclusion zones are properly planned, project-level planning can be relieved from demand questions and can be based on the assumption that the potential sites might likely have no severe negative impacts at least concerning the criteria assessed in the zoning process.

4.3 Energy Supply Systems Tailored to Spatial Structures

Energy systems design should be specific for certain spatial contexts, as can be already derived from the explanations above. This subchapter deals with principles to adapt energy supply systems to spatial structures, but also to recognize the chances that certain energy supply systems, which are embedded in a local and/or regional spatial context, might offer for further spatial development. Therefore, we suggest to deliberately aim for coevolution of spatial structures and energy supply systems.

Principle 5: Determine the demand of facilities for energy generation, distribution, and storage in light of (spatial) energy efficiency, regional resource potential, and mitigation of land use conflicts

As one of the main principles in energy planning, first energy saving potentials should be exhausted, and then, a tailor-made energy supply should be designed, meeting the demand as close as possible. Therefore, by connecting spatially contextualized energy efficiency potentials including further spatial development and regional resource potentials, the demand of additional facilities for renewable energy generation, distribution, and storage can be unerringly determined. Not only natural, technical, and economic potentials, but also spatially feasible potentials should be determined in integrated spatial and energy planning processes. These spatially feasible potentials take value decisions of the local and regional societies with respect to related land use conflicts into account. They serve as a basis for fixing the energy technology mix. Therefore, the demand question for energy supply facilities can be properly answered.

The danger for wrong decisions concerning huge energy infrastructure investments, which limit the further (financial) scope for the implementation of the energy turn, can be reduced taking the spatial contexts of the future energy supply systems into account. This also decreases the danger of risking intolerable environmental or socioeconomic impacts of "wrong" energy infrastructure projects. We propose that any new facility should only be approved if the demand of this facility can be argued by the scope of integrated spatial and energy planning.

A major source of inefficiency within an energy system is a lack of spatial contextualization of energy provision and distribution systems. This is already a problem in the current fossil energy system, when excess heat from power stations goes unused, reducing overall spatial energy efficiency. It becomes even more complex in light of the energy turn, as now the resource chain has to be taken into consideration as well: Wherever surplus energy resources are available, they have to be put to optimal use, as competition for the basic resource of a low carbon energy system, namely land, is fierce.

There is no one-size-fits-all solution for spatial contextualization of the energy system (see also Principle 6). Some considerations may however prevent false siting as well as oversizing of energy provision and distribution systems:

4.3 Energy Supply Systems Tailored to Spatial Structures

- Providing more heat than necessary adds to spatial inefficiency and has to be avoided. This means that any off-heat generated by industrial processes must be considered before additional energy provision technologies are installed.
- Using direct solar heat shall take precedence before any other supply possibility, as this heat offers the highest spatial efficiency.
- Heat demand should be the governing factor for the size of energy provision. As heat may not be distributed over long distances, any heat generated by CHP plants has to be used close by.
- Regions that identify a surplus of renewable resources by the consultation process described in Principle 3 shall also utilize these resources. All surplus renewable resources not necessary to support the local/regional energy system (taking the aspects stated above into consideration) shall be processed and exported. This helps the region to profit optimally from its natural resources and helps the "import" regions to become more "renewable." This is especially true for urban–rural interrelations, as urban areas have to rely on a supply hinterland to cover their energy and resource demand on a renewable basis.
- Storage is a costly element within an energy system. Storing electricity is the most expensive way to store energy (see Fig. 2.6, Narodoslawsky 2014) and shall be avoided as much as possible. This means that storage should preferably be applied to heat as well as to gas and other material energy carriers.
- As electricity is costly to store and the electricity grid has low storage capacity, stabilizing the grid becomes a priority task for a sustainable energy system. The general rule here is that balancing demand and supply on a regional (medium voltage) level avoids expensive measures to strengthen the grid and increases resilience. This can be achieved by decentral CHP plants that operate according to the electricity demand (and store the heat generated as a couple product) as well as by load shifting technologies such as power-to-heat and (where applicable) power-to-gas.

Principle 6: Design networks of technologies to connect energy domains and optimize the energy efficiency within specific spatial structures

Putting Principle 5 into realization requires complex and strongly interlinked energy provision and distribution technologies. The choice of technologies is however no technocratic decision. On the one hand, a network of technologies is always also a network of value chains, along which actors have to cooperate. It will define the pathways of resource utilization as well as the possibility to close material cycles. Finally, the technology network will be a key factor for economic success and innovation. On the other hand, the optimal technology network is influenced not only by resource availability and demand. External factors such as expected cost and price developments are constitutive for technology network choices, so are internal factors such as know-how, culture, and trust between actors (Stoeglehner et al. 2010).

This requires that the consultation process described in Principle 3 must enter a second stage that has to lead to a binding development strategy with a clear vision and objectives as well as alternatives to incrementally implement this master plan where options emerge because of changes in stakeholder perceptions, changes in spatial structures or new technological options. Within this consultation process, scenarios of technology networks based on Principle 5 may be used to facilitate stakeholder deliberations. In this processes, stakeholders will have to agree on the framework of internal and external factors described above. By using appropriate tools (as described in Chap. 6), consistent scenarios will be established and provide the basis for adopting a binding development strategy.

Principle 7: Organize coevolution of spatial structures and renewable energy systems

One strength of the integrated contemplation of spatial structures and renewable energy systems is the possibility to organize coevolution. Not only energy systems can be tailored to spatial structures, but also spatial structures can be adapted to the possibilities to utilize so far untapped energy sources, or consciously developed new energy technology networks. This highly complex principle can be explained on three examples: (1) existing core areas of spatial development; (2) industrials developments around nodes of different energy networks; and (3) the use of wastewater energy.

In the first example, renewable energy supplies can be tailored to core areas of spatial development as defined above. As these core areas normally already have an existing nucleus, they might be supplied, e.g., by district heat from a CHP system. According to our system, future mixed-function, appropriately dense and compact spatial developments, should be directed in these core areas, so that additional customers for a district heating (and cooling) system will be brought into areas where such systems are already provided. In this way, new developments get a renewable energy supply, and the existing renewable energy systems, a better economic base (Stoeglehner et al. 2014).

The second example deals with intersections of distribution grids of energy. Such nodes also provide potentials to change between the energy domains, or to transfer energy losses—normally as waste heat—to usable energy sources. Therefore, areas around the intersections of distribution grids should be reserved as sites for grid-overarching energy conversion technologies and their potential end users. Energy conversion plants at these intersections are a prerequisite to form a strongly interlinked holistic energy system that on the one hand provides stable supply to end users and on the other hand exploits existing resources with highest possible efficiency. These sites are moreover preferred sites for energy-intensive industries, as these industries may absorb excess heat generated by CHP units (Stoeglehner et al. 2011).

The third case deals with the utilization of wastewater energy recovery by heat exchangers and heat pumps, which might already happen at the location where sewage originates, in the canals, or in the wastewater treatment plants. We deal with the third option here, as in this option the sewage treatment process is not endangered by cooling the sewage water. In principle, two potential fields of application can be determined for energy recovered from wastewater treatment plants (Neugebauer et al. 2015): Within the settlements, district heating grids as defined above can be utilized in existing spatial structures and new development areas. Furthermore, also district cooling can be provided. As the second field of application, such energy sources can be used in agriculture and forestry: (1) for drying agricultural and forest products, such as grain, herbs, and wood; (2) heating and cooling of stables, e.g., for piglet breeding or poultry farming; (3) for aquacultures; and (4) for greenhouse production. The latter is especially interesting in countries with a low coverage rate of plant production. For instance, in Austria—depending on the kind of vegetable—coverage rates of only up to 60 % of domestic vegetable production are reached at the moment (BMLFUW 2014), so that this additional energy source for glasshouse production might allow for new and meaningful land uses in the proximity of such so far untapped energy sources.

Finally, it can be concluded that by bringing in additional compatible land uses in areas with high renewable energy provision and/or high amounts of waste energy, the energy efficiency of co-evolved spatial structures and energy systems can be considerable increased. Win-win situations can be generated that lead to a decrease of environmental pressures, the affordability of energy, and the economic feasibility of renewable resource systems. This special asset can only be achieved by integrated spatial and energy planning.

References

Baaske, W. (2013). Regional-statistic correlation of building density, mix of functions and labor-force participation rate of women. Unpublished work of STUDIA Austria.
BMLFUW – Bundesministerium für Land- und Forstwirtschaft, Umwelt und Wasserwirtschaft (2014). Nationale Strategie und Umweltrahmen - Operationelle Programme Erzeugerorganisationen, Sektor Obst und Gemüse. Arbeitspapier des Bundesministeriums für Land- und Forstwirtschaft, Umwelt und Wasserwirtschaft. https://www.ama.at/getattachment/e5d02964-acf8-4f37-a2be-57d2d525422e/06_Nationale_Strategie_2014_Maerz.pdf. Accessed 06 Jan 2016.
Dallhammer, E., & Mollay, U. (2008). Infrastrukturkosten der Siedlungserweiterung bei bestehenden Leitungsnetzen. http://www.oir.at/files2/download/projekte/Raumplanung/Infrakosten-Endbericht_20_anonym.pdf. Accessed 08 Nov 2015.
Felber, G., & Stoeglehner, G. (2014). Onshore wind energy use in spatial planning—a proposal for resolving conflicts with a dynamic safety distance approach. *Energy, Sustainability and Society, 4*(22), 1–9.
Narodoslawsky, M. (2014). Utilising bio-resources—a rational strategy for a sustainable bio-economy, ita-manuscript, Vienna. http://epub.oeaw.ac.at/ita/ita-manuscript/ita_14_02.pdf. Accessed 08 Nov 2015.
Neugebauer, G., Kretschmer, F., Kollmann, R., Narodoslawsky, M., Ertl, T., & Stoeglehner, G. (2015). Mapping thermal energy resource potentials from wastewater treatment plants. *Sustainability, 7*(10), 12988–13010.
Newman, P., & Jennings, I. (2008). *Cities as sustainable ecosystems: Principles and practices*. Washington DC: Island Press.
Stoeglehner, G., & Narodoslawsky, M. (2008). Implementing ecological footprinting in decision-making processes. *Land Use Policy, 25*, 421–431.

Stoeglehner, G., Narodoslawsky, M., Steinmüller, H., Haselsberger, B., Eder, M., Niemetz, N., et al. (2010). *INKOBA—Durchführbarkeit von nachhaltigen Energiesystemen in INKOBA Parks*. Wien: Final report.

Stoeglehner, G., Narodoslawsky, M., Steinmüller, H., Steininger, K., Weiss, M., Mitter, H., et al. (2011). *PlanVision—Visionen für eine energieoptimierte Raumplanung*. Final report. Wien.

Stoeglehner, G., Peer, V., Emrich, H., & Zeller, R. (2013). *Wohnbauförderung unter Berücksichtigung raumplanerischer Fragestellungen*. Unpublished project report.

Stoeglehner, G., Erker, S., & Neugebauer, G. (2014). *Energieraumplanung*. Materialienband. In Zusammenarbeit mit der ÖREK-Partnerschaft „Energieraumplanung". ÖROK Schriftenreihe Nr. 192. Wien: Bundesministerium für Land- und Forstwirtschaft, Umwelt und Wasserwirtschaft, Geschäftsstelle der Österreichischen Raumordnungskonferenz (ÖROK).

United Nations. (1992). AGENDA 21—United Nations Conference on Environment & Development Rio de Janerio, Brazil, 3–14 June 1992. https://sustainabledevelopment.un.org/content/documents/Agenda21.pdf. Accessed 08 Nov 2015.

United Nations (2016). Stakeholder Involvement. In: Sustainable Development Knowldege Platform. https://sustainabledevelopment.un.org/majorgroups. Accessed 06 Jan 2016.

Chapter 5
Measures for Integrated Spatial and Energy Planning

Gernot Stoeglehner, Michael Narodoslawsky, Susanna Erker, and Georg Neugebauer

Abstract The local and regional level are extremely important, not only because of potential solutions tied to regional and local spatial contexts, but also because of the multiple actors to be involved in the planning processes. In order to break down the visions for the spatial archetypes and the fields of action for everyday integrated spatial and energy planning, this chapter drafts requirements for integrated spatial and energy plans at the regional and local scale, and describes specific measures for different spatial structures.

5.1 Integrated Spatial and Energy Plans on the Regional Scale

As regulatory frameworks differ from country to country, we define the main contents an integrated spatial and energy plan should have. This subchapter is based on Stoeglehner et al. (2011, 2014) if no other sources are specified. The regional scale is very important in planning for the energy turn, as, for instance, urban areas have a supply hinterland that has to be taken into account, and certain technologies such as large-scale wind power can deliver energy far beyond the local or even regional energy demand and may have (positive and negative) regional environmental, social, and economic impacts. In order to integrate energy issues on the regional scale, we propose two bundles of measures, one related to energy-efficient spatial structures and other related to energy systems design.

5.1.1 Definition of Core Areas for Regional Spatial Development

In order to create energy-efficient spatial structures in line with field of action 4.1 "energy efficiency and spatial structures," regional core areas for spatial development have to be defined. They should comprise all land uses including central, residential, commercial, and industrial as well as open space and recreational land uses. This measure should include the following issues according to the principle of short-distance regions:

- Requirements for residential developments in multi-functional structures;
- Determination of areas with qualitatively high densification of settlement structures including minimum and maximum densities for new urban developments in regional core areas of spatial development;
- Definition of territorial scopes for supply facilities, including daily and upscale retail facilities, public social infrastructures, educational facilities, and open space recreational areas and securing their accessibility by public transport, walking, and biking for the majority of the regional population; integration of retail space in existing settlement structures;
- Determination of regional focal areas for industrial and commercial zones in close spatial contact to existing settlement structures in order to allow for short-distance accessibility and/or location on nodes of public transport;
- Concerning open space and cultural landscapes, determining the minimum available area for certain functions such as agricultural production, recreational areas, and ecological functions including ecosystem services;
- Development and provision for regional axis of energy-efficient and environmentally friendly mobility; connection of different settlement structures and core areas of regional development with each other; establishing core areas of regional development as nodes of high-performance regional public transport.

If such measures are implemented, compact settlement structures on the regional scale can be achieved which would also support renewable energy generation, as more production areas would be made available. Less sprawl means larger connected areas for agricultural production and energy generation as well as zones to protect biodiversity.

5.1.2 Integrated Resource and Energy Concepts

Within integrated spatial and energy plans, subconcepts dealing with resource provision and protection as well as visions for renewable energy supplies should be introduced. These integrated resource and energy concepts should contain the following:

- Surveys of regional resource potentials based on detailed spatial analysis; these surveys should cover energy supply potentials as well as other resource potentials and demands, spanning from food production to biomass-based industrial raw materials;

5.1 Integrated Spatial and Energy Plans on the Regional Scale

- Spatial potential analyses should include determining the tolerated level of land use and landscape transformation as well as environmental impacts generated by renewable energy generation, e.g., via exclusion areas or priority zones for different energy generation technologies such as wind energy, hydropower, open space solar power plants, and biomass energy production;
- Definition of regional energy saving and energy efficiency targets, specific for each spatial archetype in the region; the spatial archetypes (Chap. 3) can be further divided and specified to the neighborhood scale (see Sect. 5.3);
- Definition of the coverage of the total regional energy demand by regionally available renewable energy sources, taking both the spatially determined resource potentials and the spatially differentiated energy saving potentials into account; the difference between coverage and total energy demand reveals the need for energy imports into the region or the options for energy exports; at least on a global scale, in the long run, the coverage has to be 100 % if the energy turn was successful.

By introducing spatial analysis and impact assessment in energy concepts, their quality and target achievement can be substantially increased. Potential analysis, which is often based on natural, technological, and economic potentials for energy supplies, can be complemented by spatial analysis, taking land use conflicts and opposing interests in a specific region into account. Therefore, such analyses can consider real implementation potentials of energy supply technologies on specific sites, as base for planning processes and discussions between stakeholders. This makes the smooth realization of projects more likely, especially if all relevant stakeholders and actors are integrated in such planning processes (see Fig. 4.2). The planning process for such energy and resource concepts can balance interests of different groups as well as public interests like nature conservation. On the regional scale, this is extremely important if the impacts of energy generation technologies reach beyond the local scale, e.g., with large-scale wind park developments, hydropower plants, and open space solar power plants, which may also cause new demands for high-voltage power lines. How such processes can be designed is discussed in Chap. 6.

A further asset of such energy and resource concepts is the possibility to take their results in the development consent processes for energy generation, energy distribution, and energy storage facilities into account. As developers, operators, and investors normally propose concrete projects, the approval of such projects should be oriented to the planning foundations drafted above. The energy and resource concepts already answer questions like: (1) Are enough regional resources available for the operation of a facility?, (2) is the demand high enough to use the capacity?, or (3) in case not, are energy distribution networks strong enough to export the regional overproduction? In other words, they make it possible to answer the demand question for certain energy supply facilities on a strategic level. Therefore, they can be an important planning foundation for both public sector planning and private sector project development.

5.2 Integrated Spatial and Energy Plans on the Local Scale

Integrated spatial and energy plans on the local scale should be not only holistic, but also three-dimensional. Only the consideration of the third dimension makes the implementation of design principles as laid out in Chap. 4 feasible. In this section, we introduce the contents such plans should have, spanning from a spatial development strategy to integrated land use plans and building schemes, again after Stoeglehner et al. (2011, 2014). We present this section in four subchapters—building land, open space, infrastructure and mobility, and energy. The planning contents attributed to these four aspects of local spatial planning should be considered as intertwined, influencing each other, and not sectored.

5.2.1 Measures Concerning Building Land

The first question concerning building land should address the issue of feasibility from an integrated spatial and energy planning perspective. This feasibility includes the possibility to create mixed-function areas and the implementation of short-distance accessibility between different spatial functions. If certain building land uses are zoned, they should be bound to the existence of other land uses in the vicinity. For instance, zoning for retail stores, commercial areas, or public infrastructures should be connected to the availability of residential functions. When it comes to industrial areas, which might be separated because of emission protection reasons, a basic quality for grid-bound infrastructure (like transport and energy) should be met, e.g., to allow commuters to use public transport or bikeways, to have railway access at least for large-scale industrial zones, to use industrial waste energy in district heating systems, and to allow for energy storage facilities or the connection of energy domains such as electricity and heat. Measures within built areas are mainly addressed to the active system elements' mix of functions, siting, and density according to Chap. 2.

Within the building land, the first priority should be to define "core areas of spatial development" that represent urban, town, or village centers from a functional perspective, not necessarily from a historic viewpoint. These core areas should be multi-functional and at least moderately dense. Functions have to include apartments, daily supply and other shopping facilities, services, and educational, research, healthcare, and other public facilities appropriate for the level of centrality of the respective locality. Such central facilities should be concentrated in these core areas, an appropriate density of inhabitants and other functions should be guaranteed, and renewal processes should be prioritized in such areas before greenfield developments. Therefore, public subsidy programs, such as housing subsidies and economic subsidies should be focused on such core areas of spatial development. This would also increase the effectiveness of subsidy programs. Concerning the energy supply, such core areas of spatial development are interesting zones for district heating. First, the density is higher than that in other areas, multi-functionality tends to level out demand dynamics and makes grid-bound systems more efficient and economic, and different energy sources along a grid can be used, making energy cascades (see Fig. 2.3) applicable, including waste heat from industrial zones or heat recovered from sewage systems.

The choice of sites for different spatial functions defines the options to implement objectives of integrated spatial and energy planning to a large extent, not only in core areas of spatial development. Spatial developments, also in the industrial sector, should not be allowed without connection to existing settlement borders. Not only the applicability of grid-bound energy supplies, but also the feasibility of public transport, walking, and biking is related to the choice of sites. Additionally, siting can consider local conditions such as microclimate, topography, and exposition, which heavily influence energy efficiency on the project level and the active and passive usability of building-integrated solar power (see Fig. 2.4). Furthermore, brownfield developments should be prioritized before greenfield developments.

Finally, density is an issue of high relevance that has to be treated with care. As a measure of efficiency, higher densities are favorable as discussed in Sect. 2.1. But as an indicator for quality of life, there are limits to density both concerning the lower limit—too small densities make infrastructures inefficient and lead to a decrease in all issues of (daily) supply—as well as the maximum level, where too high densities lead to discontent of the inhabitants and to the desire to leave such areas. Therefore, both minimum and maximum densities for certain spatial structures should be defined, e.g., as floor space index. In order to incorporate not only energy efficiency measures, also the mix of social groups should be supported, e.g., by introducing zones for social housing apartment buildings in different spatial structures, including core areas of spatial development. Furthermore, denser areas should be located close to the nodes of public transport.

The following core measures concerning building land should be taken:

- Determination of core areas for spatial development, concentration of supply and public facilities in these areas, high-quality densification, and prioritized urban renewal in these areas;
- Siting of building land in order to guarantee for mix of functions and to optimize local spatial conditions (e.g., exposition, topography, microclimate) for future building projects;
- Give priority to brownfield developments before greenfield developments;
- Definition of minimum and maximum densities for certain spatial structures;
- Zoning areas for mixed-function houses and apartment buildings including social housing projects;
- Statements about the active and passive use of solar energy, e.g., building sizes, location of buildings to each other, exposition of living rooms, and measures to utilize microclimatic conditions.

5.2.2 Measures Concerning Open Space

Open space is an important factor for quality of life and influences the mobility demand of the population. In the long run, open space also determines the wish of people to move outside of undersupplied areas, which creates new demand for embodied energy and transport (e.g., commuting) and also often leads to the choice of low-density areas—in other words, undesired developments from the viewpoint of designing energy-efficient spatial structures. Therefore, also from the perspective of integrated spatial and energy planning, the supply with recreational areas, playgrounds, and sports fields in sufficient size, proximity, and accessibility by walking and biking is essential. Private, semiprivate, and public open space should be taken into account.

Furthermore, open spaces and their design can increase the attractiveness of biking and walking routes. A high-quality design needs space, what has to be reflected in land use planning. Therefore, sufficient space for attractively designed local routes for environmentally friendly means of transport, separated from car traffic, should be zoned in order to connect core areas of spatial development and different neighborhoods with each other and with the nodes of high-capacity public transport. This includes sufficiently large space for streets in order to allow for the separation of walking, biking, and public transport routes and car traffic as well as for green infrastructure.

Finally, open space design including vegetation and green infrastructures on buildings is an important feature to secure quality of life under climate change conditions, e.g., urban heat islands. In order to adapt to climate change, the sealing of soils should be limited and green infrastructures should be obligatorily introduced in building schemes (Pitha et al. n.y.). This might save cooling energy, and attractive living environments help to reduce individual car transport.

In a nutshell, the measures for local integrated spatial and energy plans concerning open space can be summed up as follows:

- Securing of public open spaces, recreational areas, playgrounds, and sports grounds in sufficient size, proximity, and accessibility by walking and biking;
- Securing sufficient supply with private and semiprivate open spaces such as urban gardening projects and private gardens;
- Design of car-free (residential) areas with collective garages on their fringes;
- Creation of green connections in sufficiently large road cross sections for walking and biking routes between neighborhoods, separated from car transport;
- Utilization of green infrastructures for climate change adaptation and the reduction of cooling demands.

5.2.3 *Measures Concerning Infrastructure and Mobility*

In order to reduce mobility, criteria of accessibility by walking, biking, and public transport should be introduced or strongly considered when determining whether certain areas are suitable for building land. Starting from residential functions, supply facilities and other public facilities (e.g., education, healthcare, childcare, elderly care), workplaces, and stations of high-performing public transport should be reachable in walking distance. Industrial areas should be accessible by walking, biking, or public transport.

Zoning and building schemes can guarantee that the distances for walking and biking are kept short, e.g., by high permeability of spatial structures with a dense network of walkways and bikeways and the reduction of detours between points of interest, e.g., by allowing for diagonal ways without raster orientation. Car transport may go longer ways, but people walking and biking are more sensitive to "detours" because of spatial structures. Furthermore, the number of parking spaces should be determined not only for cars (minimum and maximum values) but also for bikes. In general, mobility concepts that take the intermodal transport into account would provide a valuable input for integrated spatial and energy plans.

Finally, the measures for infrastructure and mobility can be summed up as follows:

- Definition of supply qualities concerning daily supply and public social infrastructures including maximum distances from dwellings to such facilities (e.g., kindergartens, schools, medical treatment);
- Guaranteeing a high permeability of building land for walking and biking;
- Fixing minimum and maximum parking lots for cars, specified for different spatial structures, as well as possibilities for bicycle parking;
- Definition of supply qualities of public transport, coordination of building land development, and nodes of public transport.

5.2.4 *Measures Concerning Energy Supplies*

As already stated for the regional scale, also the local-scale spatial and energy plans should contain subconcepts dealing with renewable resources and energy. On the local scale, they are addressed to the site-specific chances, potentials, and restrictions for energy efficiency and renewable energy supplies. Similar to the regional scale, local concepts should address the following topics:

- Examination and determination of the local resource base grounded on small-scale spatial analysis, taking regional provisions into account;
- Definition of local energy efficiency and energy saving targets for the respective localities and neighborhoods; development of energy efficiency and supply

standards specific for neighborhoods and urban districts applicable in renewal and expansion processes of spatial structures;
- Definition of the coverage of locally available (renewable) energy sources taking local energy efficiency and energy saving potentials into account:
  - Determination of priority areas for grid-bound energy sources, especially district heating zones and expansion zones for district heating for both combined heat and power cycles and the utilization of untapped heating and cooling potentials;
  - Examination of untapped heating and cooling potentials such as industrial waste heat, waste incineration, caloric electric power plants, and sewage energy recovery as well as biomass district heating resource potentials;
  - Determination of measures for building-integrated solar power use, e.g., definition of building-integrated solar energy recovery per dwelling, inhabitant, employee, or square meter floor space;
- Definition of decision criteria for the selection of technologies for energy generation, distribution, and storage on the local level and the possibility to exclude certain energy sources in certain areas, especially fossil fuels or other technologies with intolerable local spatial impacts like large-scale wind power;
- Clarification of the demand question for energy provision, distribution, and storage facilities on local scales;
- Securing of resource recovery areas as priority areas for certain energy generation technologies (such as wind, photovoltaics, biomass district heating) and securing of sites for the respective facilities and devices, taking the environmental and socioeconomic impacts on the neighborhoods into account.

A special feature of integrating energy concepts in spatial plans is the possibility to introduce new land uses in the vicinities of waste heat sources, as was elaborated in Sect. 4.3, Principle 7, e.g., around wastewater treatment plants (Neugebauer et al. 2015). Integrated spatial and energy planning may not only reveal such new possibilities, but it can also address them in spatial planning processes and provide the platform for discussing them with the local population.

5.3 Measures for Existing Spatial Structures

Planning in the greenfield, even the conversion of large, unused, or underused brownfields is relatively easy compared to the change in existing spatial structures that are under full use. Still, especially in developed countries, a huge building stock is already available, and spatial structures are defined which are often not in line with the ideas presented in this book, so that change is necessary, but will very likely be incremental to a large extent, as spatial structures are persistent against change.

So the questions remain, (1) what can be done in existing structures, and (2) how can we proceed toward more sustainable, energy-efficient spatial structures that can be supplied with renewable energy sources. The following subchapters are based on a discussion process between the authors and several planning practitioners, a spatial planner establishing local and regional spatial plans, a transport planner working with local mobility concepts, and three representatives of planning authorities. The results are being put together for the general public in a so far unpublished working paper in German (Stoeglehner et al. forthcoming) and are presented here for the scientific community. They cover the following archetypes of spatial structures:

- Residential areas with apartment buildings and/or mix of functions;
- Single-family housing areas;
- Mixed-function areas in central locations;
- Industrial and commercial areas;
- Shopping centers;
- Rural areas.

5.3.1 Residential Areas with Apartment Buildings and/or Mix of Functions

Summing up the criteria presented above, residential areas with apartment buildings are at least moderately dense. That means, for instance, relatively high energy efficiency, the possibility to run public infrastructures as well as daily supply shops economically efficient. It would also imply a large probability to establish economic district heating grids and, therefore, the possibility to use waste heat. Taking the discussions above into account, both minimum and maximum densities should be determined in order to provide for a high quality of life. This also has to be accompanied by qualitatively high open space provisions. Taking climate change adaptation into account, green infrastructures are of great importance in this type of spatial structures. The density values and open space requirements should be differentiated in the spatial archetypes presented in Chap. 3.

Although apartment buildings are favorable from many perspectives, including energy issues, the possibilities to provide daily supply and public infrastructures in mixed-function areas and the opportunities to design high-quality-of-life neighborhoods, many people wish to live in a single-family house (see Sect. 5.3.2). This is maybe due to the fact that many neighborhoods with apartment buildings do not provide the qualities we propose, are too dense and monofunctional, and do not have sufficient open space supply. Furthermore, qualities that are normally sought in single-family houses, such as an own garden or open space belonging to the dwelling, enough storage room, and workshops, have to be more actively and thoroughly introduced in the design of apartment buildings.

Mix of functions can be achieved in two levels: (1) in pure residential areas by an intensive network with other land uses, e.g., retail, services, or commercial areas, so that other functions can be accessed by walking or biking and (2) as mix of functions within buildings, where the lower floors can be used for retail space, public infrastructures such as kindergartens or healthcare facilities, and office space and the upper floors for dwellings. Such structures would comply with the above-described "core areas for spatial development." It also means that this densification should mainly take place in areas where multi-functionality can be organized in short distance.

Concerning mobility, no matter if areas with apartment buildings are mono- or multi-functional, a high permeability and accessibility for walking and biking should be guaranteed, and the areas should be close to nodes or at least stations of a high-performing public transport. Progressive urban design would aim at car-free areas within the settlement units and collective garages on their fringes. Furthermore, sufficient bike-parking should be guaranteed, as well as car- and bike-sharing considered.

Regarding energy supply, energy efficiency is of great importance. This includes construction technologies, but also the embodied energy in public infrastructures. As density is higher, the demand for embodied energy per capita or square meter floor space in mixed-function areas is lower than that in other settlement structures. In order to exhaust these potentials, also the design of streets should aim at resource and energy saving, e.g., by the smallest possible asphalt or concrete strips appropriate for the respective type of street. As already pointed out, the use of waste energy from different sources is a major issue in such areas, as they provide good ground for district heating both because of density and the possibility to by trend level out consumer patterns in mixed-function areas. Furthermore, coverage by building-integrated solar energy use should be determined, which also affects the patterns how buildings are organized to each other and how shadowing effects should be taken into account both from buildings or from green infrastructures.

5.3.2 Single-Family Housing Areas

Single-family houses are normally less energy efficient than areas with apartment buildings. The degree of inefficiency is dependent on the size of the land parcels. The lower energy efficiency can be reasoned as follows: First, the demand of embodied energy per dwelling is high, e.g. for technical infrastructures as roads, sewage systems etc. Second, from a mobility perspective also the distances to reach other spatial functions are longer. Normally, densities are too small to provide efficient public transport, district heating, or economic daily supply. As floor space per person is higher, also the embodied energy increases even if buildings are constructed highly efficient and would not need energy for room heating.

In single-family housing areas, certain measures to improve spatial structures should be taken into consideration when opportunities incrementally pop up, such as:

- the use of empty sites, e.g., for modest densification, e.g., several family houses with maximum three floor buildings;
- the building of semidetached houses, row houses, or terrace houses;
- the separation of single-family houses into two- or three-family houses, e.g., as multi-generational houses;
- the configuration of smaller parcels of land in order to reduce embodied energy in public infrastructures;
- in areas close to centers, long-term transformation of single-family housing areas in areas with apartment buildings.

As the density in single-family housing areas is lower, the coverage with autochthonous solar energy might be higher. When taking the daily patterns of supply and demand into account, such facilities make either energy storage or connections to grids necessary. If the coverage exceeds 40 %, the integration of photovoltaic energy requires additional grid capacity (Ramirez Camargo et al. 2015).

Concerning mobility, short distances to "core areas of spatial development" and public transport are required, as well as high permeability for walking and biking. Furthermore, the planning of streets should be oriented not only on cars, but also on stay with a high quality, as open space is mainly private. Furthermore, streets should be designed as meeting places in the absence of other possibilities to meet (e.g., shops, cafes), as densities are very probably too low for such facilities.

5.3.3 Mixed-Function Areas in Central Locations

This type of areas corresponds to the "core areas of spatial development." From the perspective of creating center structures, they should always include residential functions. Without residents, the vividness of centers cannot be guaranteed at all times including night and weekend, and therefore, also social control lacks and security issues might arise. To build certain spatial structures for just certain times of the day and week also is inefficient from an energy perspective. Therefore, we propose to assign at least 50 % of the floor space to residential functions. In a small town with a two-story town center, this would mean retail space, social infrastructures or offices in the ground floor, and dwellings in the second floor. In urban centers, e.g., with six stories, this can be translated to shops in the ground floor, office space in the second and third floors, and dwellings from floor four to six. For Austrian contexts, this would be a good mixture based on experience, and in other spatial contexts, this share should be adjusted.

The multi-functionality of the ground floor should be aimed for in urban and town centers on suitable streets with higher traffic frequencies, including not only car transport, but also walking, biking, and public transport. A precondition for such multi-functionality is a certain room height, which shall be provided in building schemes. An important issue is the mix of functions, which should intensively interlock shops, office space, cafes and restaurants, cultural and public facilities, administration, educational and research facilities, etc. An important feature should be a multi-functional public open space with a high quality of stay for different user groups and green infrastructures not only for a nice ambience, but also to reduce urban heat island effects.

From the perspective of mobility, central locations are not only the areas of origin of mobility patterns, but also important target areas. Therefore, conditions for walking, biking, and public transport have to be good, including pedestrian zones and collective parking on the fringe or in multi-story car parks, but not extensively on the streets. These issues are already extensively discussed above.

In urban and town centers, often the protection of historical buildings is an issue that from an energy perspective has to be taken into account when fixing the coverage of on-site solar power generation and determining the energy efficiency targets. Such areas are especially feasible for district heating. Therefore, in combination with other areas, district heating supply areas with a sufficient energy demand to run combined heat and power cycles and accommodate all waste heat should be aimed for.

5.3.4 Industrial and Commercial Areas

Industrial and commercial areas accommodate production sectors and are heterogeneous in size, scales, and types of companies, energy demand, mobility of persons and goods as well as their development history in certain spatial structures or detached locations. Especially, old industrial sites are often contaminated so that a conversion in other types of uses is often problematic and expensive, even if the location would be suitable for other kinds of land uses. For example, in Austria, a huge amount of zoned, but never used industrial and commercial areas is available because of local spatial development strategies that overestimate the demand. Therefore, the question is not only where to place new industrial and commercial areas, but also where to give up the zoning—which in real life never was anything else than an agriculturally used plot of land—and focus the offers for such developments on locations with "good" location factors. Integrated spatial and energy planning can contribute criteria for such discussions.

First of all, mix of functions is also important for industrial and commercial areas, even though for environmental protection reasons they might have to be separated from other, sensitive land uses. The mix of functions does not happen inside industrial and commercial areas, but they should build a close fabric with other spatial functions with the shortest possible distance because of emission protection. Each development of an industrial and commercial site should be based on a vision and strategy that prioritizes certain types of companies and prepares a

5.3 Measures for Existing Spatial Structures

fast business location if companies in line with the strategy require a site. In order to prepare such a vision and strategy, we propose to think carefully about the services an industrial and commercial area might provide for the local and regional communities. These services include the following: (1) the provision of jobs; (2) the provision of certain products; (3) the use of regional resources; (4) the reduction of local and regional environmental pressures of society; and (5) the provision of energy. In order to fulfill the latter two issues, minimum and maximum energy demands for potential companies should be defined that can be either provided from other companies in the park or distributed to near settlements as waste heat. Furthermore, the companies get a clear picture about the types of other companies they can expect in their vicinity, about the energy infrastructures, and about their possibilities to work in clusters or consider a mix of branches, e.g., preliminary production in networks of companies or business-to-business cooperation. If such strategies exist for different business parks in a municipality or on the regional scale, companies looking for a new site can be directed to the most suitable industrial or commercial development area in a region. This can not only generate benefits for the companies, but may also increase the regional innovation capacity concerning the establishing of firms (Stoeglehner et al. 2010).

Important location factors concern mobility, where the near settlements should be accessible by walking and biking. At least bigger industrial and commercial areas should be sited on nodes of transport routes, providing different modes of transport including at least railways and streets. In order to mitigate land use conflicts, industrial and commercial areas should be accessible without passing residential and mixed-function areas. Mobility has three dimensions—workforce, customers, and goods. If industrial and commercial areas are located in mixed-function areas and along public transport lines, commuting can be organized in an environmentally friendly and energy-efficient way. In order to implement this objective in practice, personal transport should be separated from truck traffic so that people do not use cars for road security reasons. Therefore, also in industrial and commercial areas, road design should be attractive for walking and biking. Ideally, in industrial and commercial areas, no or little retail is located—as shops should be sited in urban and town centers—so that little consumer traffic is to be expected. If this is not the case, mobility concepts for employees also support consumers, but would have to be complemented with measures like home delivery of goods. Such mobility and energy concepts should be part of the overall vision and strategy for the respective industrial and commercial areas.

As industrial and commercial areas often demand high amounts of electricity, the local availability might cause problems especially in rural areas, where the availability of high-voltage electricity might constitute a hard location limit to such developments. First has to be asked, if such areas would be suitable for industrial and commercial zones or if they should not be located in more centered locations—but also the rural population needs working opportunities outside agriculture if the shrinking of rural areas should be limited. In order to avoid the construction of new high-voltage grids—which is often conflict-loaded—decentralized energy supplies should be considered. Furthermore, industrial and commercial areas are often important sources of waste heat or cooling energy, which can be utilized in surrounding settlements. If all or most of the waste energy can be used outside the industrial and commercial areas, the claim to reduce the environmental pressure of society can be fulfilled. Therefore, they should be sited close to the nodes of energy grids, including district heating and electricity. In the long run, this would allow for the use of power-to-heat or power-to-gas technologies once they are broadly available.

5.3.5 Shopping Centers

Shopping centers are extremely sensitive from the viewpoint of integrated spatial and energy planning. They consume a lot of land and energy and they cause an enormous amount of individual car traffic. Furthermore, they bind buying power that is no longer available for shops in mixed-function areas and "old" central shopping streets and relocate this buying power to sites oriented to individual car transport. This also leads to a decline of mixed-function areas and spurs mono-functionality not only in the shopping areas, but also in the former centers that lack shops and face vacancies in the ground floors. People reorganize their mobility patterns and choose the car more often, as supply does no longer work with walking or biking.

The mobility of customers and employees is an important factor of the energy demand of shopping centers. Parking lots often need more ground space than the shopping center itself and consume high amounts of embodied energy for their construction. This leads to intolerable land consumption and negative impacts on landscape sceneries on the one hand, but also to high energy consumption both inside the shopping centers as well as for mobility purposes on the other hand.

Therefore, an objective of integrated spatial and energy planning would be to redevelop shopping centers and locate new shopping facilities in "core areas of spatial development" and mixed-function areas according to Sects. 5.1.1 and 5.2.1. Yet, a huge amount of retail space in shopping areas is already there. In order to describe the possibilities for redevelopment, we classify shopping centers in three types:

- Type 1—conglomerates of large single shops (without pedestrian access): Type 1 does not dispose of common strategies, and as they are not attractive to spatial functions except shopping, a transformation in mixed-function areas is not realistic. Shopping centers of Type 1 are solely car-oriented. Long distances between the single shops, the unattractive street space, and dangerous environment for pedestrians motivate customers to even change shops by car. Therefore, also public transport is not convenient. Sites often do not have a direct connection to settlements. The distances between residential areas and dislocated shopping facilities prevent to organize daily supply according to the principle of nearness. The amount of goods per purchase increases to an extent that the car becomes the most feasible means of transport. Existing areas of Type 1 should not be further expanded, but the shopping function should be brought back in center locations, except for certain groups of goods such as cars or construction material. Existing structures should be transformed to more attractive areas for biking and walking so that the connection to public transport makes more sense. Such measures are cost-intensive and have to be accompanied by measures aimed at reducing car transport, e.g., by home delivery offers and charging of parking fees in shopping centers.

- Type 2—large shopping malls with inside walking: This is a feasible model for inner-city shopping centers in cities and towns with high centrality, if such shopping malls are connected to high-performance public transport or can be accessed by walking and biking by the local inhabitants. If they are mono-functional structures outside the settlements, they should be dismissed for further development as Type 1. Inside the mall, an attractive pedestrian network is already available, so that the external accessibility by public transport reduces car dependency. Therefore, the amount of parking space can be reduced in combination with parking fees. Within urban centers, they should be combined with office space and residential functions.
- Type 3—small, suburban shopping centers: normally at ground level, shops in a row or in an L-form with a central parking lot. Such existing structures can be developed to achieve more land use and energy efficiency as well as mix of functions, at least if they are located close to existing settlement structures. First of all, the amount of floors should be expanded, e.g., by shopping space on several floors, or the combination with office space in the upper floors, further central facilities, or, if appropriate, residential use. The central parking lots should be redesigned as attractive open spaces with green infrastructures and high quality of stay. This includes reducing the amount of parking space, creating attractive walkways, and connecting these areas to public transport and local bike routes. Given a certain location quality, Type 3 can be developed as multi-functional centers mainly focused on daily supply. Such development would be favored if retail space is limited compared to other spatial functions, e.g., to 50 % in four-story buildings.

The energy side of shopping centers has different facets. Concerning electricity demand, shopping centers constitute hot spots, e.g., because of the cooling demand of shops. Because of their large (flat) roof space, they would also provide attractive sites for building-integrated photovoltaic systems. If the developers do not want to run such facilities on their own, they should be obliged to make this space available to other operators of PV systems.

Heating and cooling demands can be reduced by energy efficiency measures. Remaining demand can be covered by district heating and cooling, e.g., ran by wastewater energy recovered with heat pumps. Measures that increase energy efficiency normally increase construction costs even though running costs may very likely decrease, so that depreciation times are prolonged. Therefore, siting for these persisting spatial structures becomes even more important, as investments have to be considered long term and as structural change of such—from the viewpoint of integrated spatial and energy planning—mostly unwanted spatial structures becomes even more difficult.

5.3.6 Outer Rural Areas

Rural areas constitute a complex fabric of different spatial structures in itself. Rural small towns, market villages, and larger settlements are already included in the spatial structures described above. Farmhouses, often in dispersed locations, and scattered single-family houses without agricultural production characterize the outer rural areas. For agricultural uses, dispersed settlement structures make sense, as they combine living, working, and subsistence supply on the farm site. Yet, population working in agriculture decreases, and many farm businesses are quit with the people still living there, so that the unit of working and living is broken up. Supply and disposal and infrastructure maintenance become hard to organize and expensive, and may only be economically feasible in the larger settlements such as small towns and market villages. The questions such as which land uses are still making sense in these areas and how—if this is a desired societal objective—the local population can be motivated to stay in these areas against the global trend of urbanization arise.

Taking the high costs, and from the perspective of integrated spatial and energy planning, the high energy demand per capita into account, settlement developments should not be expanded in outer rural areas. They should be focused on the rural core areas of settlements, where supply facilities should be stabilized and maintained, which need a critical mass of people in the respective spatial structures in short distances. Only in this way, a sufficient quality of life and daily supply can be secured for the rural population. Therefore, the further development of town and village cores with mixed functions can serve as an alternative model for dispersed settlement structures. It also means to give up certain, non-agricultural structures in the long run, when buildings fall empty.

Such focused settlement structures create new space for resource recovery areas needed for the energy turn and open up new possibilities for economic activities in rural areas. By tapping renewable energy sources for electricity, heat generation and mobility (e.g., wind, solar power, biogas, biomass, biofuels, if appropriate) as well as raw materials for chemical industries (via biorefineries), new fields of economic income may emerge in pursuing the energy turn. Energy generation beyond the local demand, which means for urban and suburban demand, should be aimed for, both because of regional-economic reasons and the fulfillment of the energy turn on the global scale.

As density is low, district heating is normally not an option, so that energy-efficient houses in combination with solar thermal energy, photovoltaic-supplied heat pumps, and furnaces are feasible energy sources for room heating. Mobility is very likely oriented on cars, which leads to a high energy and resource intensity, but public transport in low-dense outer rural areas is very likely restricted by economic feasibility. In order to improve energy and climate balances for mobility, alternative fuels like electromobility have to be applied. Embodied energy is an important factor especially in rural areas. Distances are long, density is low, so that streets, sewage systems—if present—etc. pose high energy and resource demands for construction and maintenance, and are expensive. A future focus on rural core areas of spatial development would also be favorable from this viewpoint.

5.4 Concluding Remarks

Taking the network of actors and the potential fields of governmental action presented in Chap. 4 into account, it is evident that the implementation of such measures is full of barriers and pitfalls. Therefore, a consistent framework for action is necessary to successfully achieve such spatial structures, especially as they are not new to spatial planners. Many issues described above are already formulated in other visions for spatial planning (see, e.g., CNU 2001; European Commission 2011; Newman and Jennings 2008; Register 2002), even though the viewpoint of integrated spatial and energy planning adds some facets and provides additional arguments to ultimately start working on these spatial structures.

This calls for measures additional to integrated spatial and energy planning, which can be seen in the following domains: One of the main problems is the access to properties in desired core areas of spatial development. Different strategies to claim land for the implementation of new visions for spatial development have been proposed and called for, starting from the nineteenth century with the vision of Garden Cities by Howard (1965) or the Charter of Athens representing modernist urban planning visions (Hilpert 1988), just to name a few. Modest means of requiring land in public interest, e.g., for social housing and mixed-function areas, should be considered, e.g., compulsory measures to allow for certain developments, economic and fiscal measures like the direction of subsidies in core areas of spatial development, tax reliefs for desired spatial structures or higher taxes for undesired structures, development contracts between planning authorities, land owners, and developers to agree upon certain planning measures, or the establishment of public funds as players on real estate markets to buy land, develop, and resell it to the end users.

Furthermore, awareness rising for the energy turn and the issues of integrated spatial and energy planning is a precondition to find the democratic legitimation for such measures and to find actors to actually behave accordingly. Planning processes can be an important arena not only for this awareness rising, but also for the negotiation of visions, targets, and strategies of integrated spatial and energy planning. How such processes can be designed and which planning tools might facilitate such processes will be discussed in Chap. 6.

References

CNU—Congress for the New Urbanism. (2001). Charta of the New Urbanism. https://www.cnu.org/sites/default/files/charter_english.pdf. Accessed 08 Nov 2015.

European Commission (2011). Cities of Tomorrow. Challenges, Visions, Ways forward. http://ec.europa.eu/regional_policy/sources/docgener/studies/pdf/citiesoftomorrow/citiesoftomorrow_final.pdf. Accessed 08 Nov 2015.

Hilpert, T. (1988). LeCorbusiers "Charta von Athen" - Texte und Dokumente. Krit. Neuausg., 2. Edition, Braunschweig [et al]: Vieweg.

Howard, E. [1902] (1965): Garden cities of to-morrow. Edited by Osborn, F.J. MIT-Press: Cambridge, Massachusetts.

Neugebauer, G., Kretschmer, F., Kollmann, R., Narodoslawsky, M., Ertl, T., & Stoeglehner, G. (2015). Mapping Thermal Energy Resource Potentials from Wastewater Treatment Plants. *Sustainability, 7*(10), 12988–13010.

Newman, P., & Jennings, I. (2008). *Cities as Sustainable Ecosystems: Principles and Practices*. Washington DC: Island Press.

Pitha, U., Scharf, B., Enzi, V., Mursch-Radlgruber, E., Trimmel, H., Seher, W., Eder, E., Haslsteiner, J., Allabashi, R. & Oberhuber, A. (n.y.). Grüne Bauweisen für Städte der Zukunft. Optimierung des Wasser- und Lufthaushaltes urbaner Räume mittels Gründächern, Grünfassaden und versickerungsfähigen Oberflächenbefestigungen. http://www.gruenstadtklima.at/download/leitfaden_GSK.pdf. Accessed 08 Nov 2015.

Ramirez Camargo, L., Zink, R., Dorner, W., & Stoeglehner, G. (2015). Spatio-temporal modeling of roof-top photovoltaic panels from improved technical potential assessment and electricity peak load offsetting at a municipal scale. *Computers, Environment and Urban Systems, 52*, 58–69.

Register, R. (2002). *Ecocites: Building cities in balance with nature*. Berkeley California: Berkeley Hills Books.

Stoeglehner, G., Erker, S. & Neugebauer, G. (2014). *Energieraumplanung*. Materialienband. In Zusammenarbeit mit der ÖREK-Partnerschaft „Energieraumplanung". ÖROK Schriftenreihe Nr. 192. Wien: Bundesministerium für Land- und Forstwirtschaft, Umwelt und Wasserwirtschaft, Geschäftsstelle der Österreichischen Raumordnungskonferenz (ÖROK).

Stoeglehner, G., Narodoslawsky, M., Emrich, H. & Koch, H. (forthcoming). *Klimaschutz durch Raumplanung – Impulse für eine kommunale Energieraumplanung*. Working paper in preparation.

Stoeglehner, G., Narodoslawsky, M., Steinmüller, H., Steininger, K., Weiss, M., Mitter, H., et al. (2011). *PlanVision – Visionen für eine energieoptimierte Raumplanung*. Wien: Final report.

Stoeglehner, G., Narodoslawsky, M., Steinmüller, H., Haselsberger, B., Eder, M., Niemetz, N., et al. (2010). *INKOBA – Durchführbarkeit von nachhaltigen Energiesystemen in INKOBA Parks*. Wien: Final report.

Chapter 6
Processes and Tools for Integrated Spatial and Energy Planning

Gernot Stoeglehner, Michael Narodoslawsky, Susanna Erker, and Georg Neugebauer

Abstract So far, we have extensively discussed how the elements of a joint spatial and energy planning system are intertwined and how their effectiveness for proceeding toward the energy turn can be judged; we have drawn visions for integrated spatial and energy planning, described fields of action and principles as well as measures to achieve energy efficiency and renewable energy supplies in different spatial structures. This chapter is dedicated to the design of planning processes for the energy turn and tools that can facilitate such processes. First, a theory frame for process and tool design is introduced. Second, the interplay between top-down framework planning and bottom-up action planning is described. Finally, tools to support such planning processes are introduced and their applicability for certain planning tasks is discussed.

6.1 Theory Framework

The considerations of the following chapter are based on the understanding of planning as societal learning processes with the aim of strategy formation, with strategies consisting of a vision and an action plan with concrete implementation measures. Visions are bound to a spatial context, reflecting the base values and their weighting to each other, and are desirable potential futures of a certain locality (Fürst and Scholles 2001). The energy turn is a major system change with multiple effects in overcomplex systems, loaded with uncertainties about environmental and social effects, economics, and technological development. Therefore, learning has to involve the reflection of the value base expressed as a vision and measures to achieve it in light of perceived consequences of the visions and measures. These learning processes can take place on two levels (Argyris 1993; Innes and Booher 2000): single-loop learning and double-loop learning (see, Fig. 6.1). If a planning process is recognized as a three-step approach, with drafting a vision as the representation of the value base as the first step, determining an action plan with concrete measures as the second step, and appraising perceived consequences as the third step, single- and double-loop learning can be defined as follows (Stoeglehner 2010):

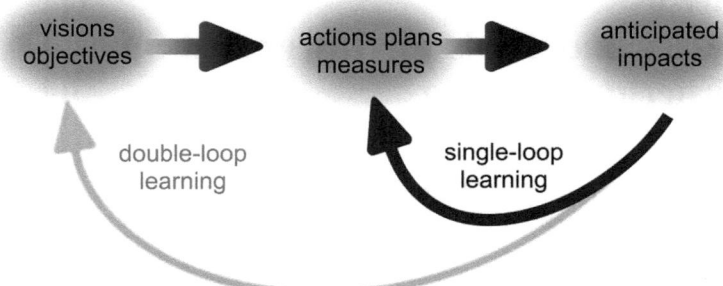

Fig. 6.1 Single- and double-loop learning (own illustration after Argyris 1993; Innes and Booher 2000; Stoeglehner 2010)

- Single-loop learning: If consequences of the vision and action plan cannot be tolerated, measures can be adapted. This can be done by changing the sites or technical layout of the proposed measures, and/or by introducing additional compensation measures for undesired negative effects. As a result of single-loop learning, the action plans will be adopted without changing the underlying visions and values of the planning process.
- Double-loop learning: This may take place if single-loop learning does not lead to acceptable perceived consequences. In this case, the adaptation of action plans is not enough, so that the visions and values of the planning process have to be challenged. Double-loop learning results in changed visions and, therefore, alternative action plans which then should only have tolerable perceived impacts.

With a model of decision making in mind, that deciding means to connect a level of facts with a level of values (Fürst and Scholles 2001), influenced by power relations and actor constellations (Scharpf 2000), single-loop learning can be determined as learning about facts and double-loop learning as learning about values. These levels of learning processes can also be pictured by three kinds of alternatives: system alternatives, site alternatives, and technological alternatives (Therivel 2006; Stoeglehner 2010). With respect to integrated spatial and energy planning, they can be described as follows:

- System alternatives: comprise alternative visions for planning solutions. Concerning integrated spatial and energy planning, they comprise different ways to achieve energy-efficient settlement structures, bundles of energy efficiency strategies, alternatives mixes of applied resources, or alternative technology mixes to achieve the energy turn, all related to a specific spatial context. Most importantly, system alternatives deal with the question of demand for further projects.
- Site alternatives: represent potential locations for certain measures in line with the system alternatives. In integrated spatial and energy planning, they represent, for instance, different locations for diverse kinds of land uses and projects in order to achieve energy-efficient spatial structures, energy efficiency measures,

location and size of priority areas for certain renewable resource recovery zones, sites for energy generation, distribution, and storage facilities.
- Technical alternatives: consist of technical solutions for certain projects on a chosen site. In integrated spatial and energy planning, it would be the concrete design of housing areas, infrastructures, energy supply facilities, etc.

The model of decision making presented so far might suggest that planning and learning processes would be linear: Visioning, action planning, and appraisal of impacts might lead to single-loop learning first, then double-loop learning; the development of alternatives might start with system alternatives on which site alternatives are based, and finally, technical alternatives which are derived from system and site alternatives. Yet, real-life planning processes are more iterative, delineating different feedback loops. For instance, it is necessary to "jump" between the types of alternatives in strategy formation, as, for instance, knockout criteria on the level of technical alternatives might hinder the implementation of a vision. This change of levels of alternatives increases the complexity of planning processes, as imponderabilities on the level of site and/or technical alternatives might make the reflection of visions necessary, as well as undesired consequences might cause the redrafting of a vision (Stoeglehner 2014). For a planning process, this means that assessments of consequences, as well as a rough estimation of measures arising from system alternatives (i.e., visions in the sense of double-loop learning), should already be carried out when a vision is drafted. If the appraisal of consequences takes place after action planning, normally too many predecisions have been taken in the course of the planning process so that the questioning of visions is no longer possible (Stoeglehner 2014). Therefore, integrated spatial and energy planning has to provide for learning processes, where (1) the preliminary assessment and adaptation of visions is made possible on rough estimations of measures; and (2) the "jumping" between the levels of alternatives can be operationalized by respective decision-making processes.

Finally, the processes of integrated spatial and energy planning have to be designed in a way that, as already stated in Chap. 4, a multitude of actors will learn about their individual behavior being in line with the energy turn. Therefore, they have to recognize that their (everyday) decisions have an impact on the achievability of the energy turn. They have to "want" the energy turn and have to understand the energy turn as something meaningful for society and for themselves in order to act accordingly. Stoeglehner et al. (2009) call this kind of dedication to planning processes "ownership." This "ownership" has three dimensions: ownership of values, ownership of processes and techniques, and ownership of outcomes.

Based on implementation theory (see, e.g., Lipsky 1980; Mazmanian and Sabatier 1983; Winther 1990), "street level bureaucrats" (SLBs) (Lipsky 1980)—the people who actually have to implement policies such as climate change mitigation or the energy turn—have to be considered as sense makers for these policies (Stoeglehner et al. 2009). By reinterpreting such policies in their everyday actions, they decide, inter alia, if a certain policy is successful, if it leads to system changes or minor adjustments. Given the multitude of actors relevant to integrated spatial

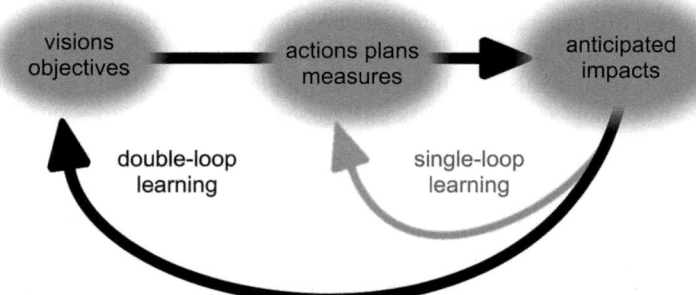

Fig. 6.2 Double-loop learning with strategic planning and assessment methods (own illustration after Stoeglehner 2014)

and energy planning (see, Fig. 6.2), many different kinds of SLBs have to develop "ownership" for the energy turn, e.g., people working in planning authorities, with developers, operators, and investors as well as end users. Ownership emerges when SLBs have fully understood the policy, have been able to participate in the visioning process with the possibility for double-loop learning, and have trust in processes and techniques applied, so that they finally want to see the outcomes of the planning process implemented. These requirements pose major challenges to planning processes, as many actors (SLBs) have to learn the same things although they have different interests, realities, and living worlds. In the following subsections, we present our approach to deal with these complexities of process design.

6.2 Top-Down Framework Planning and Bottom-up Action Planning

The drafting of visions and action plans for the energy turn is often based on potential analyses, definition of objectives, and suggestions of action plans as science-driven processes. Such rational approaches place scientific survey in the center of planning, work with base values legitimized by democratic powers, and design a master plan to reach the objectives. They are heavily criticized, as conclusive information is never present, as planning is a value full activity (Lawrence 2000), and as rational approaches tend to disguise normative aspects as scientific knowledge. Furthermore can be observed that the societal base values concerning the energy turn are not sufficiently negotiated, e.g., by controversial positions toward energy turn effects such as the food versus fuels debate or landscape change by biofuel production, wind power generation, and large-scale PV-installations as well as huge controversies about the expansion of high-voltage grids in Germany and Austria, which are presently planned and reasoned as a precondition to integrate renewable electricity generation into the energy system.

Although the main aim to protect the climate by changing the energy supply systems to a renewable resource base is widely accepted, the normative base that allows for the judgments of technological options is not yet agreed on. Some of the innovative potential studies involve stakeholders to estimate the social acceptance of changes in the energy system, but if certain energy projects can be realized in a certain regional or local context heavily depends on the interests present in the respective area, the actors involved and their opinions brought forward in the planning process. This corresponds to the communicative and collaborative planning paradigm that puts debate about the options for future developments in the center of planning and focuses on consensus between decision makers, planners, and the public (Healey 1992; Müller 2004). Yet, this approach is criticized, inter alia, for the problem that consensus is unlikely in conflict-driven situations—as can be observed around huge energy infrastructure projects—and that not all relevant interests for a planning process can be present. For instance, sustainability calls for the considerations of interests of future generations, which, by definition, cannot participate in a planning process (Fischer 2003). In order to overcome such gaps between rational decision and communicative planning, Stoeglehner (2010) proposed to combine the rational and communicative planning paradigms: To make a decision, the level of facts has to be integrated with the level of values, as already stated above. In order to reach this integration, certain rules of aggregation (which can be translated as planning methods) have to be applied. In the rational–communicative paradigm, it is proposed to found the normative base—the negotiation and agreement of values—and the agreement on aggregation rules (i.e., planning methods) on collaborative debate, whereas the level of facts and the application of the aggregation rules can be a scientifically grounded exercise. Interests not present should be represented by advocacy, for instance by relevant authorities, NGOs or the planners themselves.

In order to reach an efficient process of stakeholder involvement, we suggest the combination of top-down framework planning and bottom-up action planning (Stoeglehner et al. 2010): In our experience, in such overcomplex problem settings as integrated spatial and energy planning, it is difficult for decision makers to enter a public debate without a clear picture and the possibility to have an own, well-argued opinion and position as a starting point for discussions. The top-down framework planning, which involves planners and decision makers, serves as a rough determination of potentially feasible planning options that can be supported and argued by the decision makers, including their socioeconomic and environmental impacts. The result of the top-down framework planning is a set of planning options that are justifiable for the decision makers and compatible with the overall development of the respective planning scale (national/regional/local). The top-down framework planning consists of eight steps (modified after Stoeglehner et al. 2010):

1. Definition of objectives: taking development objectives for the planning area, including social, economic, and environmental objectives as well as objectives for spatial development, resource utilization and energy supply into account;

2. Survey of energy demand and resources: energy and resource demand of the respective societies and economies as well as resource potentials in the planning area;
3. Spatial analysis: analysis of potentials, challenges and location factors of spatial development and estimation of the future energy demand based on potential spatial developments;
4. Site analysis: analysis of existing and potential sites concerning energy supplies (energy provision and distribution) taking their potential for certain supply options and their environmental effects into account;
5. Framework definition for scenario building: with respect to available resources considering potential land use conflicts as well as technological, societal, and economic boundary conditions for scenarios;
6. Development of scenarios: for technologies and technology networks for optimized energy and resource supplies, taking demand, resources, technologies, and products into account;
7. Scenario assessment: using general methods for environmental (e.g., cumulative energy demand, ecological footprints, CO_2 life cycle emissions), social (e.g., jobs created), and regional–economic (e.g., regional income generated) impact appraisal;
8. Selection of feasible options: as the basic starting point for the bottom-up action planning.

This model is challenging for decision makers as the results of the top-down framework planning have to be debated from the scratch with local and regional stakeholders, actors needed for the implementation of measures and the affected public. Phase 2 not only serves as validation of assumptions made in phase 1, but means a discussion of the renewable energy future of a certain area with the target groups of plan implementation. All actors involved shall have the same information than the decision makers from the expert-driven steps—survey of energy demand and resources, spatial analyses, and site analyses. Therefore, the result has to be open and may not be prejudiced by the decision makers within the given framework. If no solution is agreeable, even the framework planning might have to be repeated. The bottom-up action planning includes 7 steps (modified after Stoeglehner et al. 2010):

1. Selection of actor groups: All groups of actors have to be represented, either who are important for the realization of renewable energy supply projects or affected by them, or consumers of energy who have to reduce the demand or should be convinced to choose certain supply options;
2. Reflection of the objectives: discussion and, if necessary, amendment of the objectives;
3. Calibration of resources and demand: It has to be checked whether the estimated potentials are really available, e.g., if house owners are willing to install solar energy facilities on their roofs, if farmers are willing to dedicate certain areas for biomass energy production;

6.2 Top-Down Framework Planning ...

4. Reflection of scenario frameworks: discussion and, if necessary, amendment of the scenario frameworks, e.g., in light of social acceptability of technologies, energy prices, environmental impacts;
5. Scenario development: repetition of scenario development with the reflected and amended boundary conditions;
6. Scenario assessment: of the new scenarios with the same methods than top-down framework planning;
7. Selection of agreed option: agreeing on planning options and start of the implementation of the approved action plan.

By implementing such planning processes, we take up that feasible energy futures can be agreed on by a multitude of stakeholders and actors (SLBs) relevant for a successful implementation. The broad consensus is necessary as such a complex problem as the energy turn needs "ownership" and action of almost everybody—may it be a change in consumer behavior, the engagement in energy saving, the installations of renewable energy devices, or just the assurance of societal support of energy turn measures. This is especially important as the system and actor analyses revealed that the energy sector alone very likely will not be capable of achieving the energy turn as this not only requires supply-side measures, but also demand-side measures. Furthermore, some of the renewable energy provision technologies can be operated on a small scale, which means that consumers are empowered to become prosumers—so that they consume and provide energy at the same time. This makes decentralized approaches to energy supply regulation necessary and adds to the social, economic, and organizational complexity of future energy supply systems. Moreover, most of the system elements concerning energy supply are passive, especially when it comes to demand. Therefore, trying to change the energy system only within the energy sector would not be very efficient as it means to use the wrong leverage. This calls for an integrated approach toward spatial and energy planning that has to be reflected not only by the design of planning processes, but also by the tools applied.

6.3 Tools for Integrated Spatial and Energy Planning

Tools have two main tasks in planning processes: (1) They should facilitate learning about fact and values; and (2) they should reduce the information load on decision makers. In integrated spatial and energy planning, where the value base for decisions and the visions are still not clear, double-loop learning is of high importance. In order to facilitate double-loop learning in planning processes, methods and tools are needed that allow to recognize consequences of visions with only little information about action plans, in other words support shortcuts to double-loop learning. Anticipated consequences of planning options can already be appraised when choices at the system level become evident and if rough assumptions about related measures and actions in order to implement a possible set of targets are made. Such planning and assessment methods can be called "strategic" (Stoeglehner 2014).

Strategic planning and assessment methods allow for a preassessment on the level of system alternatives, use few indicators, draw a general picture about potential impacts, and provide for an "unsustainability test" according to the indicator pyramid (Fig. 6.3, Stoeglehner and Narodoslawsky 2008): The indicator pyramid shows that in the course of a decision process, the information load on planners, decision makers, and involved actors increases.

On top of the pyramid, relatively early in the decision-making process, system alternatives are developed. If they are appraised with strategic planning and assessment methods, this unsustainability test can be carried out. In the indicator pyramid, only planning options on the system level which can be identified as potentially sustainable can be advanced and elaborated by considering constant further appraisal. Suitable indicators for environment-related preassessments at the system level are, for instance, energy footprints, CO_2 life cycle emissions, or cumulated energy demand, in socioeconomic domains, for instance, regional revenue, import–export balances, regional jobs created, etc. (see, e.g., Stoeglehner et al. 2011a, 2014b).

Such strategic planning and assessment methods are in itself complex, as they reduce overcomplex real-life systems to models that have to be traceable and understandable to multiple groups of actors with diverse backgrounds. By allowing these planning actors—including planners and decision makers—to enter double-loop learning, they can negotiate an agreed value base, a vision, and the respective action plans for its implementation. These learning processes can be especially fruitful, if the changes in visions and action plans and their related consequences can be modeled fast. In the best case, planning actors experience this in workshop situations, so that the planning actors can adjust their planning measures and can immediately see how these adaptations or the introduction of

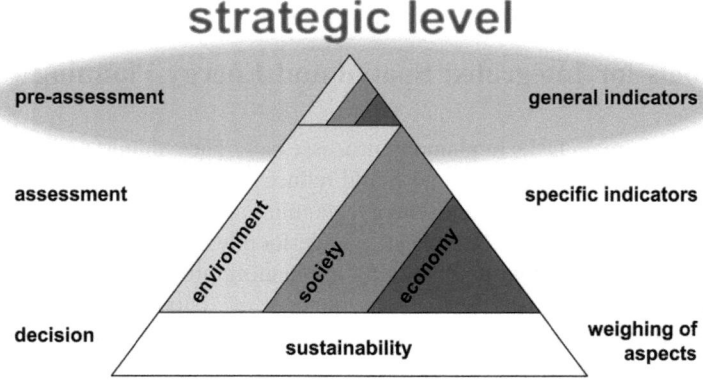

Fig. 6.3 The indicator pyramid (own illustration after Stoeglehner and Narodoslawsky 2008; Stöglehner 2014)

compensation measures might change the appraisal of consequences. Therefore, strategic planning and assessment methods work best if they are transferred to related planning tools.

In order to provide an insight into this topic, four tools will be presented. These tools chosen have been codeveloped by at least one member of the author team of this book and were already applied in planning cases. Each tool introduction includes the main objectives, a basic description of the functionality, a short review of potentials and limitations, and a section about the specific kinds of learning processes and learning outcomes that can be achieved with each tool. The tools selected are (1) the ELAS calculator which was designed as the first full implementation of the indicator pyramid; (2) Energy Zone Mapping, a tool to design energy supplies taking spatial dimensions into account; (3) Energy Pass for Settlements 2.0, an energy performance rating for settlement structures; and (4) RegiOpt, a tool to optimize regional resource flows.

6.3.1 Elas

The ELAS calculator is an online tool which can be used for long-term energy analysis of existing or planned residential developments. The aim of ELAS is to characterize planned or existing housing estates from an overall perspective, while taking into account spatial planning criteria for energy and associated environmental and socioeconomic impacts. ELAS enables its users to evaluate, modify, and optimize settlements or single objects based on simple inputs from an energetic, environmental, and socioeconomic point of view. Therefore, the calculator illustrates the interlinkages of energy consumption, energy supply, siting of settlement structures, and mobility.

In detail, the tool generates a set of four different groups of results: (1) the energy consumption resulting from heating demand, hot water preparation, electricity, mobility of residents, and the energy demand of infrastructure; furthermore, embodied energy of buildings and infrastructure is calculated, if new constructions take place; (2) CO_2 life cycle emissions; (3) the ecological footprint; and (4) regional economic effects such as revenue, value added, imports, and jobs. Figure 6.4 illustrates one part of the result section.

Based on this, the establishment of settlements as well as the renovation, expansion, demolition, and reconstruction of residential buildings can be simulated and evaluated. Both in the case of a status analysis and planning deliberations, modifications can be made by varying the input parameters. ELAS provides predefined scenarios which show the effects of spatial development on a long-term basis. At first, there is the possibility to form a "Trend Scenario," based on current forecasts and trends of spatial development. The second predefined scenario is called "Green Scenario," which alleges a more sustainable use of energy and

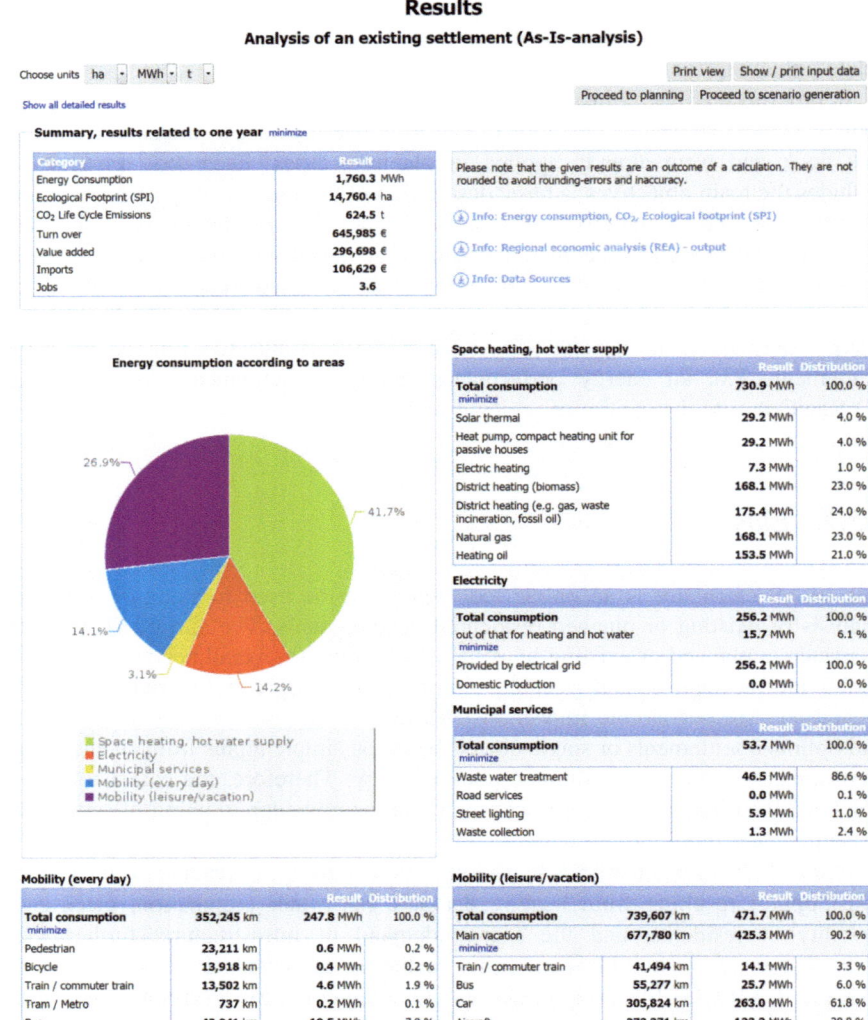

Fig. 6.4 The energy demand of a residential settlement as one part of the ELAS results (Stoeglehner et al. 2011a)

resources in the future. Finally, ELAS provides an evaluation of planning alternatives by comparing different locations, building structures, technological choices with respect to the construction of buildings, forms of energy supply, variations in household sizes, and changes in mobility behavior.

The ELAS calculator is a Web-based freeware tool in English and German, see: www.elas-calcualtor.eu. It provides results for various target groups such as planning community, architects, developers, and interested actors. Accordingly, the calculator analyzes settlements or individual buildings from different angles and enables its users to draw conclusions for different problems and levels of decision making.

By using the ELAS calculator, the energy consumption and the environmental and socioeconomic consequences of different settlement developments can be estimated. The range of possible developments is produced by changing input parameters, applying given scenarios or comparing different variants. Thus, ELAS provides a basis for discussion and decision making in order to optimize residential developments from an energetic and environmental perspective.

Despite the large amount of data behind the ELAS calculations, additional estimations were necessary. Consequently, assumptions based on own surveys, questionnaires, studies, and statistics had to be implemented in the tool. A probability of error cannot be provided; however, most of the assumptions within ELAS are kept transparent and changeable in order to avoid sources of error. For this purpose, the tool suggests default values which can be adapted if more accurate information is provided by the user. This flexible calculation leads to adaptive options for action.

ELAS builds the basis for decision making regarding residential settlements. It helps to create an overall understanding for energy-efficient, environment-friendly, and climate-protecting conditions when designing residential settlement structures.

6.3.2 Energy Pass for Settlements 2.0

Nowadays, energy performance certificates for buildings are mandatory in Austria. Due to these instruments, energy efficiency standards of buildings can be visualized and optimized. However, this does not ensure energy-efficient solutions for settlements, since these current energy certificates for buildings ignore the spatial context. In this case, a zero-energy house on the "greenfield" gets a positive evaluation due to its eco-friendly construction standards, even though the mobility of the inhabitants or the embodied energy in infrastructures may be high because of a peripheral location. This circumstance is not justifiable from an integrated spatial and energy planning perspective, since the aim is to enable overall optimization, not just for individual units but whole residential areas. From this point of view, additional criteria should be considered such as settlement structures and density, location of buildings, site development, and the availability of technical and social infrastructure.

Following the idea of the already-established energy certificates, the "Energy Pass for Settlements 2.0" was created by using Microsoft Excel in order to strengthen settlement developments considering aspects of integrated spatial and energy planning. Here, the "Energy Pass for Settlements 2.0" pursues the goal of evaluating existing or future residential settlement structures in terms of energy efficiency, the reduction of greenhouse gas emissions, and the minimization costs by applying the principles "density," "compactness," "short distances," and "multi-functionality." The overall objective is to achieve high quality of living within residential settlements, while creating more sustainable spatial structures.

Based on defined, categorized, and classified parameters, different variants of spatial development as well as diverse locations can be compared to each other. Hence, the tool takes qualitative and quantitative data into account to ensure a comprehensive consideration of relevant topics. In order to start the evaluation, site-specific inputs are required concerning "general information," "local conditions," "accessibility and mobility," "external infrastructure," and "additional investments." Furthermore, project-specific queries are implemented into the tool focusing on "project descriptions," "environmental quality," and "internal infrastructure." On the site and project level, the tool provides a detailed result section with two main parts and four dimensions of results: a benchmarking of qualitative inputs as well as an evaluation of energy consumption, costs, and CO_2-emissions (see, Fig. 6.5).

The first part of this section constitutes the evaluation of the "choice of location." Therein, a benchmarking of framework conditions to ensure social and technical infrastructure gets visualized. Hereinafter, the energy consumption as well as the costs of spatial development is getting estimated, followed by an evaluation of the CO_2 emissions caused by the estimated mobility behavior of additional residential population. In the end of this section, all aspects of location are summarized. The second part of this section is dedicated to the building structure and the configuration of the chosen location. In the beginning, information about overall characteristics is getting summarized and the structure of the project area is shown. Afterwards, the energy demand and costs to set up and maintain these settlement structures as well as the environmental quality are evaluated.

Summarizing these results, the tool provides an overall evaluation of settlement structures, energy consumption, costs as well as CO_2 emissions based on the concept of energy efficiency classes (A–G) in order to assess residential settlement structures. Similar to the energy certificates and established strategies of environmental planning (Fürst and Scholles 2001), all parameters as well as their linkages got benchmarked and rated in the course of a year-long process of discussion between tool developers. Thereby, "A" stands for the most energy-efficient use of resources and spatial structures, while "G" represents the worst achievable rating.

Additional to this final benchmarking, a comparison of building development scenarios can be found in the end of the tool, which gives a quick overview of the impacts of different development strategies on population and building densities.

6.3 Tools for Integrated Spatial …

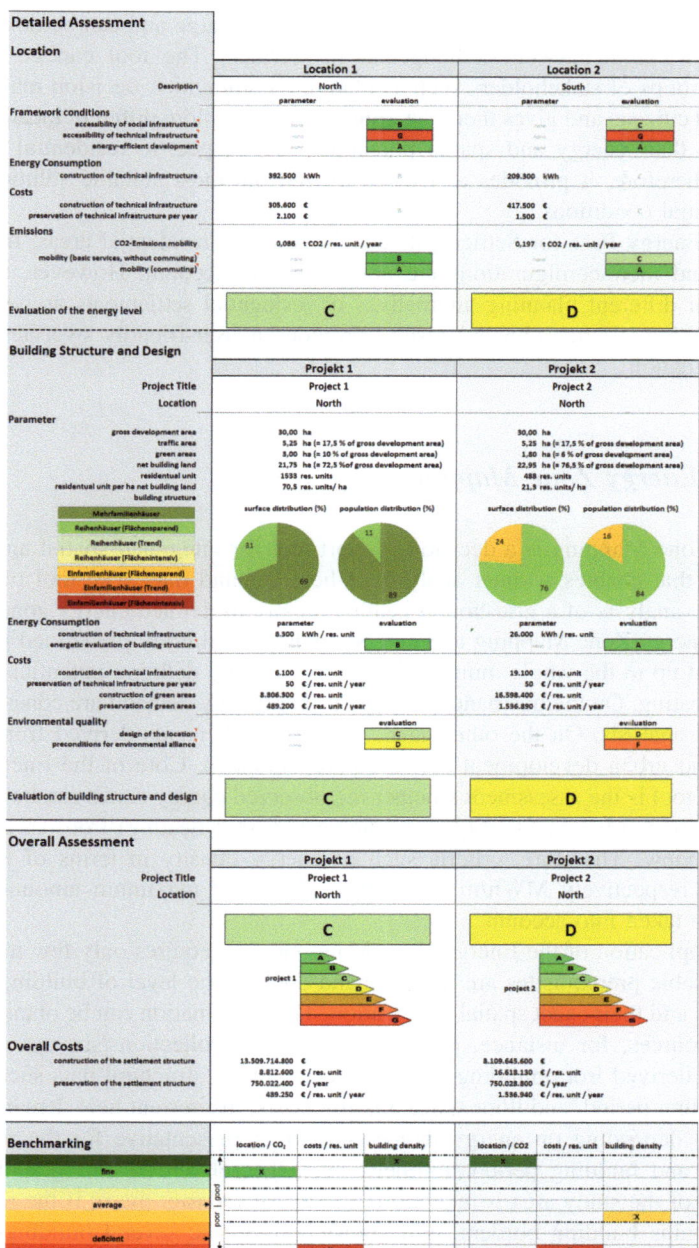

Fig. 6.5 The final rating of the Energy Pass 2.0 (Emrich et al. 2012)

The "Energy Pass for Settlements 2.0" is free of charge and can be downloaded from www.energieausweis-siedlungen.at in German. The tool can be used by various groups of stakeholders such as planning community, decision makers, and interested citizens and gives them the opportunity to explore different locations with regard to their energy and spatial potentials with respect to residential developments. Therefore, it provides assistance to find the most suitable solution under given spatial conditions.

The "Energy Pass for Settlements 2.0" focuses on residential areas. Individual objects and their configurations are not taken into account. However, the comparison of different planning alternatives of residential settlements as well as the benchmarking of the achieved results provides a user-friendly overview of the spatial situation.

6.3.3 Energy Zone Mapping

Energy Zone Mapping is a decision support tool for integrated spatial and energy planning that enables a zonal analysis of heat demand data. The tool provides a feasibility analysis of a grid-bound heating supply in defined energy zones.

The Energy Zone Mapping tool can be applied to arbitrary predefined parts of a settlement up to the whole municipal territory, e.g., by defining potential areas for district heating. On the one hand, data on current energy demand are considered for the zonal analysis. On the other hand, future heat demands derived from energy saving and urban development scenarios are estimated. Core of the Energy Zone Mapping tool is the assessment whether the observed energy zones can be provided with grid-bound heating supply (e.g., biomass district heating) from an economic point of view. Therefore, criteria such as energy density in terms of heat load (kWh/m, respectively MWh/m) as well as a defined maximum amount of heat losses are taken into account.

The application of the Energy Zone Mapping tool requires only few input data. Indispensable prerequisites are heat demand data on the level of buildings and/or addresses and their exact spatial localization. This information can be obtained from several sources, for instance, energy demand data collections/surveys or if not available derived from building inventories containing structural data such as land use, building period, and floor space. On this basis, the current heat demand can be estimated depending on energy indices that are representative for the respective land use and building period. Furthermore, a georeferenced cartographic representation of the study area is required (e.g., digital cadaster map). If the respective map contains building borders, grid connections to the several buildings can be modeled exactly resulting in a high quality of the related estimations. Furthermore, the aggregation of several parcels to microdemand sections enables a simple and clear presentation of the respective results. In addition to this basic data, the

6.3 Tools for Integrated Spatial ...

generation of two new data sets is required containing, firstly, a hypothetical district heating grid, and secondly, the respective house connections. Helpful, but not mandatory for the creation of these data sets can be the utilization of orthophotos.

Figure 6.6 illustrates the results of a zonal analysis of the heat demand for an Austrian urban settlement considering eight energy zones (Stöglehner et al. 2011b). The left illustration depicts the current energy demand calculated on the basis of energy indices derived from building periods. In a second step, future heat demands are estimated (central illustration) as it can be assumed that future heat demand will change, for instance, through the implementation of thermal insulation or the expansion of floor space of buildings.

The Energy Zone Mapping tool offers two heat demand scenarios: The first scenario constitutes a moderate energy saving and efficiency potential calculated with a 20 percent reduction of the current heat demand. The second scenario estimates the future heat demand applying adopted energy indices that represent a realistic and achievable target status for each building type and building period after upgrading the building stock's energy performance. Finally, energy densities in terms of heat load (right illustration) as prerequisites for a cost-efficient operation of a district heating system are calculated for both the current and the future heat demand scenarios. In the Energy Zone Mapping results window, an overview of the grid parameters is displayed including grid length, total heat demand, heat load, and heat losses as a share of the total heat demand. These results allow for the appraisal, whether a grid-bound heat supply in the observed energy zones can be provided from an economic point of view.

The Energy Zone Mapping tool was developed in the framework of the research project PlanVision (Stoeglehner et al. 2011b). The tool comprises a collection of macros in the scripting language VBA (Visual Basic for Applications) and was

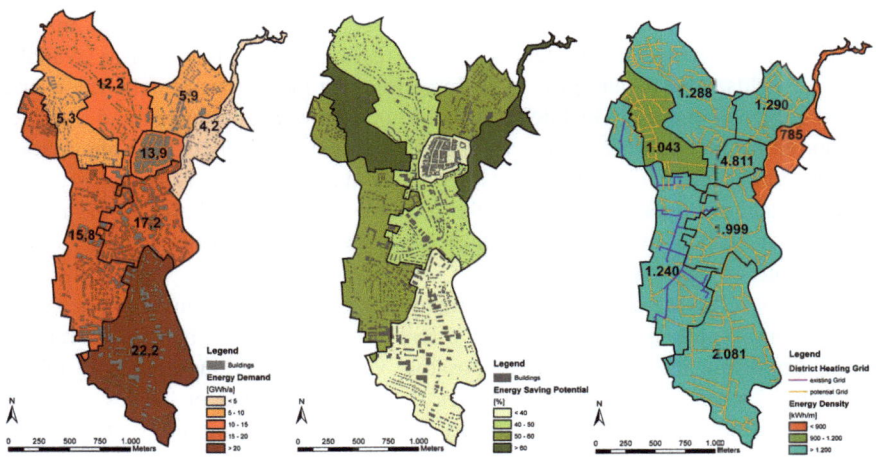

Fig. 6.6 Energy Zone Mapping as a basis for the local development concept of an Austrian urban settlement (own illustration after Stoeglehner et al. 2011b)

embedded in the software ArcGIS. The method of the Energy Zone Mapping tool is documented in Stöglehner et al. (2011a, b). On the basis of this information, the method can be implemented by well-trained GIS specialists whereby expert knowledge is necessary for the replication.

Applying the Energy Zone Mapping tool can lead to interesting learning options for several groups of stakeholders such as planning community, developers, energy suppliers, and network operators. In Freistadt, for instance, the optimal district heating supply system—based on the results of the Energy Zone Mapping application illustrated in Fig. 6.6—served as a starting point for the definition of district heating priority and supply areas. These measures were enacted in the municipality's local development concept and a second biomass district heating plant was established.

6.3.4 RegiOpt

RegiOpt allows the user to develop optimal technology networks utilizing regional renewable resources from solar radiation to wind to biomass to satisfy regional energy demand. It applies process network synthesis (PNS) (Halasz et al. 2005) based on bipartite graph representations of material and energy flow networks (Friedler et al. 1995) to generate an economically optimal technology network linking regional resources with regional demands. This optimal technology network will subsequently be ecologically compared to the current situation by the Sustainable Process Index (SPI) (Krotscheck and Narodoslawsky 1996), a kind of ecological footprint and the Carbon Footprint.

RegiOpt takes the form of an online questionnaire. The user is asked to provide data about the region in question regarding its population, building standard, and rough estimates on the life style (e.g., average meat consumption and mileage travelled by passenger cars operated by the regions citizens) as well as available land for fields, forests, and grassland. In addition to these general data, the user is asked for existing energy technologies based on renewable resources and livestock supported by farms in the region. RegiOpt automatically deducts land requirement for providing nutrition to inhabitants, livestock support, and existing energy technologies and provides the user with the resource base still available in the region. The user may then decide how much of these resources should be used as a basis for a future energy system.

RegiOpt offers the option to regionalize yields of solar radiation, crops as well as production and transport costs. It also offers a wide variety of help tools, ranging from supporting information about possible values for various entries to ethical hints, e.g., when a region already relies heavily on imports or causes extensive ecological pressure outside its boundaries. Every entry is supplied with a meaningful default value, allowing the user to run the tool even on less than complete data.

6.3 Tools for Integrated Spatial …

RegiOpt provides the user with a first rough estimate of how regional renewable resources can be best utilized to make the energy system more sustainable. The result pages detail the technologies within the optimal energy system, providing information about their size, resource requirement, and costs as well as flow network from resources to and in-between technologies and finally to end use. The optimal network will be compared to business as usual in ecological and economic terms. Figure 6.7 shows a screenshot of the ecological comparison, detailing the ecological pressure of nutrition, heat and electricity provision, and mobility as well. The bars also identify of how much of this ecological pressure is incurred inside and

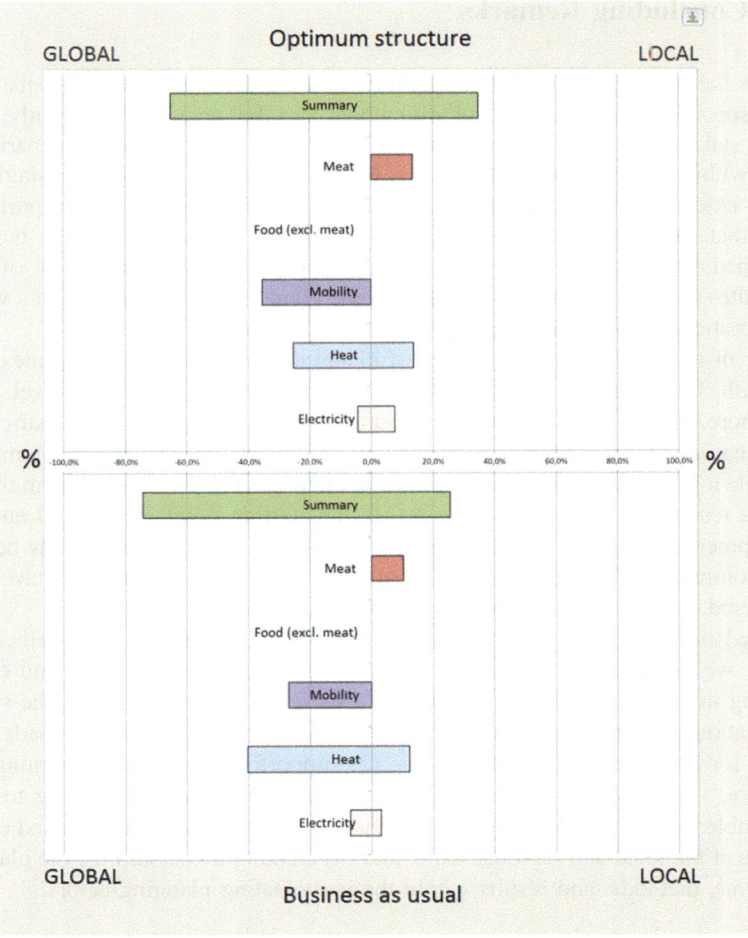

Fig. 6.7 Ecological comparison of an optimal structure with business as usual, using the SPI—ecological footprint (own illustration after RegiOpt 2016)

outside of a region. The results also compare overall economic value added, SPI and Carbon Footprint of the optimal resource utilization with business as usual.

RegiOpt is a free online tool (available from www.fussabdrucksrechner.at) that intends to help regional decision makers to obtain an overview on their development options. The tool can be used to generate scenarios by altering not only resource availability, but also life style and economic boundary conditions. A particular option for scenario generation is the possibility to limit available capital, so that users may be able to prioritize particular technological pathways that provide the largest benefit for the (scarce) money available.

6.4 Concluding Remarks

Besides the overall positive learning effects and the decision-supporting quality of tools, specific characteristics and differences can be revealed. Inherently, tools enable stakeholders to compare different results, planning variants, scenarios, or trends within a predefined methodology. In this regard, the insights of a single tool can be extended by combining several tools. For this purpose, it is important to know that the results of different tools are never fully congruent as they are developed from various data sets and assumptions. The vast diversity of tools constitutes their added value, since they enable planning actors to carry out a variety of assessments from different perspectives (Stoeglehner et al. 2014a).

One of the distinguishing features of planning tools is the spatial context they deal with. Only a few tools focus on the regional and the municipal level, while even more refer to districts and settlement structures. Most tools are restricted to individual locations or objects. Generally, it can be ascertained that the number of available tools increases in case they refer to the project level. This is due to the fact that the residential function usually makes the starting point for method and tool development. Moreover, this distribution is based on the fact that models become more complex with an increase in spatial scale, since more parameters have to be considered, coordinated, and implemented (Stoeglehner et al. 2014a).

Based on our experiences with method and tool development as well as tool testing, we propose that planning tools can support integrated spatial and energy planning in four dimensions: (1) creating a basic understanding about the system interrelations of integrated spatial and energy planning, which can be made operational for concrete planning processes; (2) supporting double-loop learning and, therefore, visioning for integrated spatial and energy plans; (3) leading to more sustainable decisions about the planning and design of spatial structures and energy systems at the local and regional scale; and (4) creating ownership for the planning objectives, methods, and results within the participating planning actors.

References

Argyris, C. (1993). *Knowledge for action: A guide to overcoming barriers to institutional change.* San Francisco: Jossey Bass.
Emirch, H., Zeller, R., Stoeglehner, G., & Erker, S. (2012). *Energy pass for settlements 2.0. funded by the Government of Lower Austria.* St. Pölten: Department of Spatial Planning and Regional Policy.
Fischer, T. (2003). Strategic environmental assessment in postmodern times. *Environmental Impact Assessment Review, 23*, 155–170.
Friedler, F., Varga, J. B., & Fan, L. T. (1995). Decision-mapping: a tool for consistent and complete decisions in process synthesis. *Chemical Engineering Science, 50*, 1755–1768.
Fürst, D., & Scholles, F. (2001). *Handbuch Theorien+Methoden der Raum- und Umweltplanung.* Dortmund: Dortmunder Vertrieb für Bau- und Planungsliteratur.
Halasz, L., Povoden, G., & Narodoslawsky, M. (2005). Sustainable processes synthesis for renewable resources. *Resources, Conservation and Recycling, 44*, 293–307.
Healey, P. (1992). Planning through debate. The communicative turn in planning theory. *Town Planning Review, 63*, 143–162.
Innes, J., & Booher, D. E. (2000). Collaborative Dialogue as a Policy Making Strategy. Institute of Urban and Rural Development, University of California, Berkeley—Working Paper Series, Accessed Nov 08, 2015 from http://escholarship.org/uc/item/8523r5zt#page-7.
Krotscheck, C., & Narodoslawsky, M. (1996). The sustainable process index—a new dimension in ecological evaluation. *Ecological Engineering, 6*(4), 241–258.
Lawrence, D. P. (2000). Planning theories and environmental impact assessment. *Environmental Impact Assessment Review, 20*, 607–625.
Lipsky, M. (1980). *Street-Level Bureaucracy: Dilemmas of the Individual in Public Services.* New York: Russell Sage Foundation.
Mazmanian, D. A., & Sabatier, P. A. (1983). *Implementation and public policy.* Glenview: Scott, Foresman.
Müller, S. (2004). Internationale Einflüsse auf die Planungstheoriedebatte in Deutschland nach 1945 oder die Perspektiven der Planungsdemokratie. In U. Altrock, S. Günther, S. Hunning, & D. Peter (Eds.), *Perspektiven der Planungstheorie* (pp. 123–140). Berlin: Leue.
RegiOpt (2016). Optimization and evaluation of regional technology networks. Accessed Nov 08, 2015 from www.fussabdrucksrechner.at.
Scharpf, F. (2000). *Interaktionsformen. Akteurszentrierter Institutionalismus in der Politikforschung.* Opladen: Leske+Budrich.
Stoeglehner, G. (2010). Enhancing SEA effectiveness: lessons learnt from Austrian experiences in spatial planning. *Impact Assessment and Project Appraisal, 28*(3), 217–231.
Stoeglehner, G. (2014). SUP-Qualität im Planungsalltag—Überlegungen zur Planung- und Prüfmethodik. *UVP-Report, 28*(3+4), 107–112.
Stoeglehner, G., Baaske, W., Mitter, H., Niemetz, N., Kettl, K. H., Weiss, M., et al. (2014b). Sustainability appraisal of residential energy demand and supply—a life cycle approach including heating, electricity, embodied energy and mobility. *Energy, Sustainability and Society, 4*(24), 1–13.
Stoeglehner, G., Brown, A. L., & Kornov, L. (2009). SEA and planning: 'ownership' of SEA by the planners is the key to its effectiveness. *Impact Assessment and Project Appraisal, 27*(2), 111–120.
Stoeglehner, G., Erker, S., & Neugebauer, G. (2014a). *Tools für Energieraumplanung.* Ein Handbuch für deren Auswahl und Anwendung im Planungsprozess. Bundesministerium für Land- und Forstwirtschaft, Umwelt und Wasserwirtschaft (Ed.). Wien.
Stoeglehner, G., & Narodoslawsky, M. (2008). Implementing ecological footprinting in decision-making processes. *Land Use Policy, 25*, 421–431.
Stoeglehner, G., Narodoslawsky, M., Baaske, W., Mitter, H., Weiss, M., Neugebauer, G. C., et al. (2011a). *ELAS—Energetische Langzeitanalysen von Siedlungsstrukturen.* Wien: Final report.

Stoeglehner, G., Narodoslawsky, M., Steinmüller, H., Haselsberger, B., Eder, M., Niemetz, N., et al. (2010). *INKOBA—Durchführbarkeit von nachhaltigen Energiesystemen in INKOBA Parks*. Wien: Final report.
Stoeglehner, G., Narodoslawsky, M., Steinmüller, H., Steininger, K., Weiss, M., Mitter, H., et al. (2011b). *PlanVision—Visionen für eine energieoptimierte Raumplanung*. Wien: Final report.
Therivel, R. (2006). *Strategic environmental assessment in action*. London: Earthscan.
Winther, S. (1990). Integrating Implementation Research. In D. J. Palumbo & D. J. Calista (Eds.), *Implementation and the policy process opening up the black box* (pp. 19–38). New York: Greenwood Press.

Chapter 7
Résumé

Gernot Stoeglehner, Michael Narodoslawsky, Susanna Erker, and Georg Neugebauer

Abstract This chapter sums up key findings of the previous chapters. First, we reason why integrated spatial and energy planning is a mayor field of action for implementing climate protection and the energy turn. Second, a rough generic model of an energy-optimized spatial structure is sketched. Third, the process dimension and the issue of methods for integrated spatial and energy planning are addressed. Finally, we point out that the principles, measures, and tools introduced in this book will allow for multiple win-win situations between the energy turn and climate protection with a multitude of other goals of society.

The climate change challenge and the rising awareness of limitations to fossil resource availability conspire to turn a new page with regard to spatial planning. How we utilize space and how we demand energy are inextricably linked. Spatial structures and the way we organize our life in its spatial context define to a great extend if we can meet the challenges of climate change mitigation and the energy turn.

From the vantage point of planning, the intimate link between energy, spatial structures, and life styles generates a kind of *energy-space-continuum* that can only be tackled by holistic planning approaches. Overcoming the limitations and negative sustainability balance of fossil resources (as well as the risk of nuclear energy) requires that within the twenty-first century, we have to complete a profound resource reorientation, relying on renewable resources as well as on efficiency gains to cover our need for energy services. This energy turn entails an equally radical turn in the spatial organization of land uses, infrastructure, and societal activities, calling for integrated spatial and energy planning.

This turn is profound because it results from a paradigmatic change in the energy system that supports all societal activities. Parallel to the advance of fossil energy starting at the end of the nineteenth century, a decoupling of the emerging energy provision system and its spatial context can be observed: Energy sources are point sources such as mines, oil, and gas fields, often far from the areas of energy consumption. The logistical system to provide energy resources became interregional and increasingly continental, even global. Big centralized energy conversion technologies became the norm. The spatial aspects of the energy system were reduced to the optimal shape of distribution grids, linking central conversion sites with consumers.

Changing to renewable resources now means changing to decentral and dispersed sources with lower energy density and transportability as well as more volatile spatial and temporal availability, either based on direct and indirect solar energy or other environmental energy sources like geothermal heat. This means that land (and to a lesser extent sea) becomes the primary energy source. Energy provision directly competes with other ways to utilize land, like providing food and environmental services as well as preserving the variety of life forms.

At the same time, the other main resource for covering our energy services, namely energy efficiency, is strongly dependent on spatial structures and the way society organizes itself within its spatial context. Dense and functionally mixed settlements reduce the demand for primary energy to provide comfort and support a vibrant economy. Meaningful industrial clusters reduce the demand of energy required to generate economic profit.

Finally, the logistics to collect, condition, and transport energy sources from their decentral source to consumers is no longer a simple "point source–point use" system. Spatial resource availability, existing technical and transport infrastructure as well as factors like available workforce and economic potential of a region define the framework of tightly interlinked value chains and associated logistical systems for renewable energy sources. Transport grids become more complex as there is no clear distinction anymore between energy providers and consumers, as any building may be equipped with technologies to harvest renewable energy while at the same time asking for energy services.

A greater reliance on renewable resources has, however, direct consequences on landscapes and cityscapes. Harnessing wind power will alter the landscape as more turbines and possibly more high voltage electrical lines will be erected. PV and solar collectors will change the appearance of buildings and cities. Energy crops and short rotation plantations might give a new face to cultural landscapes, as will energy conversion sites like biogas plants. This means that nobody can escape the consequences of the energy turn.

As we pointed out, despite the overall systems complexity involved, only relatively few design principles have to be applied to guarantee for suitable spatial structures to support climate protection and the energy turn. These are compact, moderately dense, mixed-function spatial structures organized according to the principle of nearness. Mixed-function core areas of cities and towns as well as industrial sites should be located on nodes of energy and transport grids, making it possible to connect different energy domains and modes of transport.

Figure 7.1 shows a generic scheme of an energy-optimized city, small town, or village. The scheme depicts compact spatial structures similar to models representing urban planning visions such as EcoCities, New Urbanism, or European Cities: Around the mixed-function, moderately dense core areas, residential areas, commercial and industrial zones are arranged. The mix of functions in core areas should comprise residential functions, retail, gastronomy, cultural and educational

7 Résumé

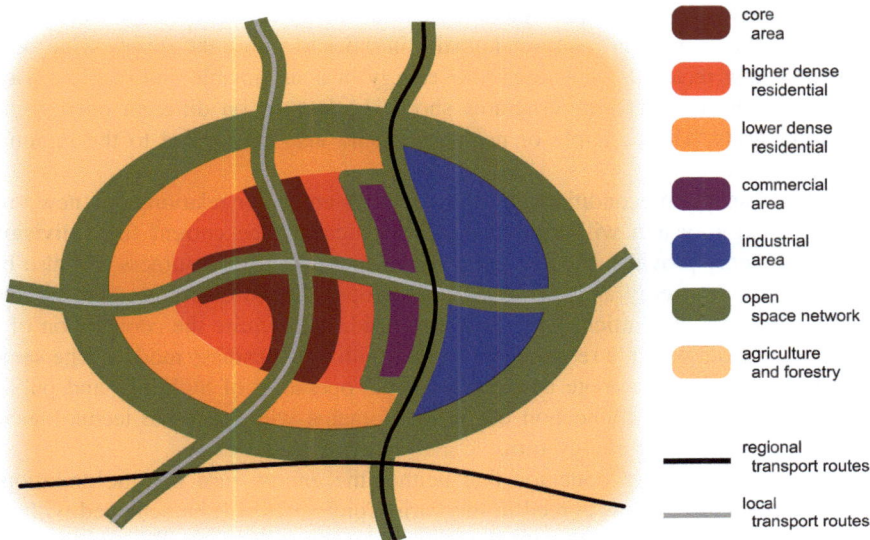

Fig. 7.1 A generic scheme of energy-optimized local spatial structures (own illustration)

facilities, public services, offices, and other functions that do not impact each other negatively. Where appropriate, mix of functions should also be organized in different floors within buildings. Open spaces are sufficiently provided both within the respective areas and for the creation of an open space grid, inter alia providing for attractive connections between the different settlement quarters for walking and biking. Between the spatial archetypes urban area, (rural and suburban) small town, suburban area, and rural area, core areas differ in the diversity of functions introduced, e.g., the groups of goods and services to be offered, and also in density. This basic spatial design shall provide for structural energy efficiency as pointed out in the previous Chapters as it allows connecting waste heat sources with heat sinks, provides for more full-load hours in grid-bound energy supplies, and opens up spatial preconditions for efficient technological choices for the energy supply.

It is likely that when considering all aspect of energy demand including residential, commercial, and industrial activities, the provision of heat, electricity, and mobility as well as embodied energy in spatial structures, the energy intensity of the described local energy-efficient spatial structures is higher than the solar harvest provided on their land covered. Therefore, only a share of the energy needed can be provided within these spatial structures. Simultaneously, because of the characteristics of renewable resources, the energy turn will make long-distance energy transport less likely. Therefore, the regional scale becomes of outstanding importance as a closer spatial and organizational connection between energy providing and energy consuming areas will become imperative. On the regional scale, the fabric of cities, rural, and suburban small towns as well as rural and suburban areas has to closely cooperate to provide for structural energy efficiency, the area required as well as the energy and resource logistics in order to secure the supply with food, industrial raw materials, and renewable energy in a sustainable and resilient way. Therefore, the local spatial structures should be located on different energy and transport grids. The principle of nearness should also be applied to the regional scale.

Besides spatial preconditions, business models and actor relations in a new low carbon energy system will also change dramatically. The current fixed division between energy provider and consumer becomes murky, as consumers will also be able to provide energy and may rely on grid operators to distribute their energy. Stabilizing grids, in particular the electricity grid, will require the cooperation of a variety of actors within a restructured market with new business models. The same will hold for energy storage as well as for the operation of buildings and public infrastructure, as the connection of energy domains will constitute technological options to support the energy turn.

From our work, we conclude that neither the energy turn nor the ambitious climate protection targets agreed in the Paris Conference will be achieved without holistic and participatory, integrated spatial and energy planning. The current book has discussed some aspects to be considered as well as measures to be taken within these planning processes. It has, however, also pointed to the fact that such a holistic planning process will need very different qualities and arrangements compared to present spatial planning.

A first difference to current planning processes is a much more level actor's field. As everyone is concerned by the spatial connotations of the energy turn, the planning process will have to become much more participatory. There is no one-fits-all solution of how the best adapted optimal energy system has to look like. Besides that, each and every actor will bring his or her interests and visions to the table. This results in a transition of the planning process toward a societal learning process, with a discourse about different possible development pathways represented by scenarios. At the end of the day, this process will have to bring together the actors in a dependable and encompassing social contract about how to shape the future in a sustainable and resilient way. Within this contract, the question of how to use the available space and how to structure it will be a central tenet.

This change in the quality of the planning process, however, requires new support tools. These tools will have to shape scenarios and evaluate the impact of decisions that have long-lasting effects on the development path of nations, regions, or cities. Although the objective of most of these tools is "only" to generate scenarios, they have to be more encompassing and arguably more precise than the tools used so far. As the decision making becomes more complex, scenarios and their environmental, social, and economic impacts will become the basis of the discourse between actors within the planning process. Any planning process, however, is only as good and solid as its factual and value base. Therefore, the quality requirements for the tools used in these planning processes have to be exceptionally high, both in terms of their scientific background and accuracy and their potential to support participatory planning. Spatial planning not only can contribute data and planning methods that can be used for designing the energy turn, but can also provide the systemic approach as well as the platforms and instruments for participatory planning that lead to legally binding results.

We are aware that existing settlement structures often develop in opposite, low-dense, mono-functional, land and energy consuming directions than the spatial structures we suggest to support the energy turn and climate protection. We are also aware that spatial structures are persistent over time. Therefore, it is all the more important to change course as soon as possible and to incrementally move in the right direction when options emerge. This book explains the systemic interrelations between spatial and energy planning and offers principles, methods, and tools to integrate these planning domains. Furthermore, different other fields of policy influence the scope of action integrated spatial and energy planning has, like housing subsidies, taxation, and the possibilities to shape real estate markets by making land available for implementing these visions. In order to achieve a consistent change in frameworks of spatial development, much awareness of the public and the respective decision makers is important. Therefore, this book also offers planning methods and tools to influence such processes of societal learning about the interrelations of spatial structures, climate protection, and the energy turn.

It is the core argument of this book that the necessity of the energy turn does not only revolutionize technology, but also the way society organizes and how it uses its spatial resources. The ideas, measures, and tools presented in this book provide a first step in this direction. They will help to create integrated spatial and energy planning processes to the requirements of the future, considering climate protection and the energy turn. Simultaneously, the suggestions presented will create multiple win-win situations with enhancing quality of life, further aspects of environmental protection, the optimization of resource use as well as the provision of green jobs and economic prosperity based on the sustainable use of natural resources. Only putting the plans into action will, however, make all the difference for the climate as well as for society.

MIX
Papier aus verantwortungsvollen Quellen
Paper from responsible sources
FSC® C105338

If you have any concerns about our products,
you can contact us on
ProductSafety@springernature.com

In case Publisher is established outside the EU,
the EU authorized representative is:
**Springer Nature Customer Service Center GmbH
Europaplatz 3, 69115 Heidelberg, Germany**

Printed by Libri Plureos GmbH
in Hamburg, Germany